A CONCISE INTRODUCTION TO THE THEORY OF INTEGRATION

SERIES IN PURE MATHEMATICS

Editor: C C Hsiung

Associate Editors: S S Chern, S Kobayashi, I Satake, Y-T Siu, W-T Wu
and M Yamaguti

Part I. Monographs and Textbooks

Part II. Lecture Notes

Series in Pure Mathematics — Volume 12

A CONCISE INTRODUCTION TO THE THEORY OF INTEGRATION

Daniel W Stroock
Department of Mathematics
Massachusetts Institute of Technology

World Scientific
Singapore • New Jersey • London • Hong Kong

Published by

World Scientific Publishing Co. Pte. Ltd.
P O Box 128, Farrer Road, Singapore 9128
USA office: 687 Hartwell Street, Teaneck, NJ 07666
UK office: 73 Lynton Mead, Totteridge, London N20 8DH

Library of Congress Cataloging-in-Publication data is available.

**A CONCISE INTRODUCTION TO THE
THEORY OF INTEGRATION**

ISBN 981-02-0145-1

Printed in Singapore by JBW Printers & Binders Pte. Ltd.

PREFACE

This little book is the outgrowth of a one semester course which I have taught for each of the past four years at M.I.T. Although this class used to be one of the standard courses taken by essentially every first year graduate student of mathematics, in recent years (at least in those when I was the instructor), the clientele has shifted from first year graduate students of mathematics to more advanced graduate students in other disciplines. In fact, the majority of my students have been from departments of engineering (especially electrical engineering) and most of the rest have been economists. Whether this state of affairs is a reflection on my teaching, the increased importance of mathematical analysis in other disciplines, the superior undergraduate preparation of students coming to M.I.T in mathematics, or simply the lack of enthusiasm that these students have for analysis, I have preferred not to examine too closely. On the other hand, the situation did force me to do a certain amount of thinking about what constitutes an appropriate course for a group of non-mathematicians who are courageous (foolish?) enough to sign up for an introduction to integration theory offered by the department of mathematics. In particular, I had to figure out what to do about that vast body of material which, in standard mathematics offerings, is said to be "assumed to have been covered in your advanced calculus course." Aspiring young mathematicians seldom challenge even the most ridiculous claims of this sort: the good ones look it up, and the others trust that "it will not appear on the exam." On the other hand, students who are not heading into mathematics are less easily shamed into accepting such claims; in fact, as I soon discovered, many of them were attending my course for the express purpose learning what mathematicians call "advanced calculus."

In view of the preceding comments about the origins of this text, it should come as no surprise that the contents of this book are somewhat different from that of many modern introductions to measure theory. Indeed, I believe that nothing has been done here "in complete generality!" On the other hand, greater space than usual has been given to the properties of LEBESGUE's measure on $\mathbb{R}^N$. In particular, the whole of Chapter IV is devoted to the applications of LEBESGUE's measure to topics which are customarily "assumed to have been covered in your advanced calculus course." As a consequence, what has emerged is a kind of hybrid in which both modern integration theory and

advanced calculus are represented. Because none of the many existing books
on integration theory contained preceisely the mix for which I was looking, I
decided to add my own version of the subject to the long list of books for the
next guy to reject.

Daniel W. Stroock
Cambridge, MA
January 1, 1990

CONTENTS

A CONCISE INTRODUCTION TO THE THEORY OF INTEGRATION

Chapter I. The Classical Theory

I.1. Riemann Integration.

We begin by recalling a few of the basic facts about the integration theory which is usually introduced in advanced calculus. We do so not only for purposes of later comparison with the modern theory but also because it is the theory with which most computations are actually performed.

Let $N \in \mathbb{Z}^+$ (throughout $\mathbb{Z}^+$ will denote the positive integers). A **rectangle** in $\mathbb{R}^N$ is a subset I of $\mathbb{R}^N$ which can be written as the cartesian product $\prod_1^N [a_k, b_k]$ of closed intervals $[a_k, b_k]$, where it is assumed that $a_k \leq b_k$ for each $1 \leq k \leq N$. If I is such a rectangle, we call the numbers

$$\text{diam}(I) \equiv \left(\sum_1^N (b_k - a_k)^2 \right)^{1/2} \quad \text{and} \quad \text{vol}(I) \equiv \prod_{k=1}^N (b_k - a_k)$$

the **diameter** and the **volume** of I, respectively. Given a collection $\mathcal{C}$, we will say that $\mathcal{C}$ is **non-overlapping** if distinct elements of $\mathcal{C}$ have disjoint interiors. The following *obvious* fact is surprisingly difficult to prove.

I.1.1 Lemma. *Let $\mathcal{C}$ be a non-overlapping finite collection of rectangles all of which are contained in the rectangle J. Then $\text{vol}(J) \geq \sum_{I \in \mathcal{C}} \text{vol}(I)$. On the other hand, if $\mathcal{C}$ is any finite collection of rectangles and J is a rectangle which is covered by $\mathcal{C}$ (i.e., $J \subseteq \bigcup \mathcal{C}$), then $\text{vol}(J) \leq \sum_{I \in \mathcal{C}} \text{vol}(I)$.*

PROOF: Without loss in generality, we will assume, throughout, that all the rectangles involved have non-empty interiors and that $I \cap J \neq \emptyset$ for any $I \in \mathcal{C}$. Also, we will number the elements of $\mathcal{C}$ so that $\mathcal{C} = \{I_1, \ldots, I_n\}$.

We begin with the case when $N = 1$. Thus, $J = [a, b]$ and each $I_\mu = [a_\mu, b_\mu]$ for some $a < b$ and $a_\mu < b_\mu$. Next, choose $c_1 \leq \cdots \leq c_{2n}$ so that $\{c_1, \ldots, c_{2n}\} = \{a_1, \ldots, a_n\} \cup \{b_1, \ldots, b_n\}$. For each $1 \leq \nu \leq 2n - 1$ and $1 \leq \mu \leq n - 1$, set

$$\mathcal{M}(\nu) = \{\mu : a_\mu \leq c_\nu < b_\mu\} \quad \text{and} \quad \mathcal{N}(\mu) = \{\nu : a_\mu \leq c_\nu < b_\mu\}.$$

When $\mathcal{C}$ is non-overlapping and $\bigcup \mathcal{C} \subseteq J$, one has that $a \leq c_1 \leq c_{2n} \leq b$ and that $\text{card}\big(\mathcal{M}(\nu)\big) \leq 1$ for each ν. Hence, in this case:

$$\sum_{\mu=1}^{n} \text{vol}\,(I_\mu) = \sum_{\mu=1}^{n}(b_\mu - a_\mu) = \sum_{\mu=1}^{n}\sum_{\nu\in\mathcal{N}(\mu)}(c_{\nu+1} - c_\nu)$$

$$= \sum_{\nu=1}^{2n-1}(c_{\nu+1} - c_\nu)\text{card}\big(\mathcal{M}(\nu)\big)$$

$$\leq \sum_{\nu=1}^{2n-1}(c_{\nu+1} - c_\nu) \leq b - a = \text{vol}\,(J).$$

On the other hand, if one simply knows that $J \subseteq \bigcup_1^n I_\mu$, then $c_1 \leq a < b \leq c_{2n}$ and $\text{card}\big(\mathcal{M}(\nu)\big) \geq 1$ for each $1 \leq \nu \leq 2n - 1$ with $c_\nu < c_{2n}$; and so:

$$\sum_{\mu=1}^{n} \text{vol}\,(I_\mu) = \sum_{\mu=1}^{n}(b_\mu - a_\mu) = \sum_{\mu=1}^{n}\sum_{\nu\in\mathcal{N}(\mu)}(c_{\nu+1} - c_\nu)$$

$$= \sum_{\nu=1}^{2n-1}(c_{\nu+1} - c_\nu)\text{card}\big(\mathcal{M}(\nu)\big)$$

$$\geq \sum_{\nu=1}^{2n-1}(c_{\nu+1} - c_\nu) \geq b - a = \text{vol}\,(J).$$

Thus, the case when $N = 1$ is complete.

To handle $N \geq 2$, we work by induction. Write $J = [a,b] \times \hat{J}$ and $I_\mu = [a_\mu, b_\mu] \times \hat{I}_\mu$, where $\hat{J}$ and the $\hat{I}_\mu$'s are rectangles in $\mathbb{R}^{N-1}$ having non-empty interiors. Next, choose $\{c_1, \ldots, c_{2n}\}$ and define $\mathcal{M}(\nu)$ and $\mathcal{N}(\mu)$ accordingly, as in the case $N = 1$. When we assume only that $J \subseteq \bigcup_1^n I_\mu$, we have

$$c_1 \leq a < b \leq c_{2n} \quad \text{and} \quad \hat{J} \subseteq \bigcup_{\mu\in\mathcal{M}(\nu)} \hat{I}_\mu \quad \text{for each } 1 \leq \nu \leq 2n - 1.$$

Hence, by induction hypothesis, in this case:

$$\sum_{\mu=1}^{n} \text{vol}\,(I_\mu) = \sum_{\mu=1}^{n}(b_\mu - a_\mu)\text{vol}\,(\hat{I}_\mu)$$

$$= \sum_{\mu=1}^{n}\text{vol}\,(\hat{I}_\mu) \sum_{\nu\in\mathcal{N}(\mu)}(c_{\nu+1} - c_\nu)$$

$$= \sum_{\nu=1}^{2n-1}(c_{\nu+1} - c_\nu)\sum_{\mu\in\mathcal{M}(\nu)}\text{vol}\,(\hat{I}_\mu)$$

$$\geq \text{vol}\,(\hat{J})\sum_{\nu=1}^{2n-1}(c_{\nu+1} - c_\nu) \geq (b-a)\text{vol}\,(\hat{J}) = \text{vol}\,(J).$$

To handle the non-overlapping case when $\bigcup_1^n I_\mu \subseteq J$, first assume that the I_μ's themselves are mutually disjoint. Then, by induction hypothesis,

$$\sum_{\mu \in \mathcal{M}(\nu)} \mathrm{vol}\,(\hat{I}_\mu) \leq \mathrm{vol}\,(\hat{J})$$

for each $1 \leq \nu \leq 2n - 1$; and so

$$\sum_{\mu=1}^n \mathrm{vol}\,(I_\mu) = \sum_{\mu=1}^n (b_\mu - a_\mu)\mathrm{vol}\,(\hat{I}_\mu)$$

$$= \sum_{\mu=1}^n \mathrm{vol}\,(\hat{I}_\mu) \sum_{\nu \in \mathcal{N}(\mu)} (c_{\nu+1} - c_\nu)$$

$$= \sum_{\nu=1}^{2n-1} (c_{\nu+1} - c_\nu) \sum_{\mu \in \mathcal{M}(\nu)} \mathrm{vol}\,(\hat{I}_\mu)$$

$$\leq \mathrm{vol}\,(\hat{J}) \sum_{\nu=1}^{2n-1} (c_{\nu+1} - c_\nu) \leq (b - a)\mathrm{vol}\,(\hat{J}) = \mathrm{vol}\,(J).$$

Finally, note that when the I_μ's are non-overlapping but not necessarily disjoint, they can be made disjoint by an arbitrarily small diminution of their sides. Hence, we can first make the necessary diminution and then pass to a limit; thereby handling the case of general non-overlapping I_μ's. ∎

Given a collection $\mathcal{C}$ of rectangles I, we will use $\Xi(\mathcal{C})$ to denote the set of all maps $\xi : \mathcal{C} \longrightarrow \bigcup \mathcal{C}$ such that $\xi(I) \in I$ for each $I \in \mathcal{C}$. Given a finite collection $\mathcal{C}$, an element $\xi \in \Xi(\mathcal{C})$, and a function $f : \bigcup \mathcal{C} \longrightarrow \mathbb{R}$, we define the **Riemann sum of f over $\mathcal{C}$ relative to ξ** to be the number

$$\text{(I.1.2)} \qquad \mathcal{R}(f; \mathcal{C}, \xi) \equiv \sum_{I \in \mathcal{C}} f(\xi(I))\mathrm{vol}\,(I)$$

Finally, if J is a rectangle and $f : J \longrightarrow \mathbb{R}$ is a function, we will say that f is **Riemann integrable on J** if there is a number $A \in \mathbb{R}$ with the property that for all $\epsilon > 0$ there is a $\delta > 0$ such that

$$\text{(I.1.3)} \qquad |\mathcal{R}(f; \mathcal{C}, \xi) - A| < \epsilon$$

whenever $\xi \in \Xi(\mathcal{C})$ and $\mathcal{C}$ is a non-overlapping, finite, **exact cover** of J (i.e., $J = \bigcup \mathcal{C}$) whose **mesh size**

$$\|\mathcal{C}\| \equiv \max \left\{ \mathrm{diam}(I) : I \in \mathcal{C} \right\} < \delta.$$

When f is RIEMANN integrable on J, we will call the associated number A in (I.1.3) the **Riemann integral of f on J** and we will use

$$(R) \int_J f(x)\, dx$$

to denote A.

It is a relatively simple matter to see that any $f \in C(J)$ (the space of continuous real-valued functions on J) is RIEMANN integrable on J. However, in order to determine when more general functions are RIEMANN integrable, it is useful to introduce the **Riemann upper sum**

$$\mathcal{U}(f;\mathcal{C}) \equiv \sum_{I \in \mathcal{C}} \sup_{x \in I} f(x)\mathrm{vol}\,(I)$$

and the **Riemann lower sum**

$$\mathcal{L}(f;\mathcal{C}) \equiv \sum_{I \in \mathcal{C}} \inf_{x \in I} f(x)\mathrm{vol}\,(I).$$

Clearly, one always has

$$\mathcal{L}(f;\mathcal{C}) \leq \mathcal{R}(f;\mathcal{C},\xi) \leq \mathcal{U}(f;\mathcal{C})$$

for any $\mathcal{C}$ and $\xi \in \Xi(\mathcal{C})$. Also, it is clear that f is RIEMANN integrable if and only if

$$\varliminf_{\|\mathcal{C}\| \to 0} \mathcal{L}(f;\mathcal{C}) \geq \varlimsup_{\|\mathcal{C}\| \to 0} \mathcal{U}(f;\mathcal{C})$$

where the limits are taken over non-overlapping, finite, exact covers of J. What we want to show now is that the preceding can be replaced by the condition

$$(I.1.4) \qquad\qquad \sup_{\mathcal{C}} \mathcal{L}(f;\mathcal{C}) \geq \inf_{\mathcal{C}} \mathcal{U}(f;\mathcal{C})$$

where $\mathcal{C}$'s run over all non-overlapping, finite, exact covers of J.

To this end, we partially order the covers $\mathcal{C}$ by *refinement*. That is, we say that $\mathcal{C}_2$ is **more refined** than $\mathcal{C}_1$ and write $\mathcal{C}_1 \leq \mathcal{C}_2$ if for every $I_2 \in \mathcal{C}_2$ there is an $I_1 \in \mathcal{C}_1$ such that $I_2 \subseteq I_1$. Note that for every pair $\mathcal{C}_1$ and $\mathcal{C}_2$ the *least common refinement* $\mathcal{C}_1 \vee \mathcal{C}_2$ is given by:

$$\mathcal{C}_1 \vee \mathcal{C}_2 = \{I_1 \cap I_2 : I_1 \in \mathcal{C}_1, I_2 \in \mathcal{C}_2, \text{ and } I_1^\circ \cap I_2^\circ \neq \emptyset\}.$$

(Here, and throughout, S° will be used to denote the interior of the set S.)

I.1.5 Lemma. *For any pair of non-overlapping, finite, exact covers C_1 and C_2 of J and any function $f : J \longrightarrow \mathbb{R}$, $\mathcal{L}(f; C_1) \leq \mathcal{U}(f; C_2)$. Moreover, if $C_1 \leq C_2$, then $\mathcal{L}(f; C_1) \leq \mathcal{L}(f; C_2)$ and $\mathcal{U}(f; C_1) \geq \mathcal{U}(f; C_2)$.*

PROOF: We begin by proving the second statement. Noting that

$$(I.1.6) \qquad \mathcal{L}(f; C) = -\mathcal{U}(-f; C),$$

we see that it suffices to check that $\mathcal{U}(f; C_1) \geq \mathcal{U}(f; C_2)$ if $C_1 \leq C_2$. But, for each $I_1 \in C_1$,

$$\sup_{x \in I_1} f(x) \operatorname{vol}(I_1) \geq \sum_{\{I_2 \in C_2 : I_1^\circ \cap I_2^\circ \neq \emptyset\}} \sup_{x \in I_2} f(x) \operatorname{vol}(I_2),$$

where we have used Lemma I.1.1 to see that

$$\operatorname{vol}(I_1) = \sum_{\{I_2 \in C_2 : I_1^\circ \cap I_2^\circ \neq \emptyset\}} \operatorname{vol}(I_2).$$

After summing the above over $I_1 \in C_1$, one arrives at the required result.

Given the preceding, the rest is immediate. Namely, for any C_1 and C_2,

$$\mathcal{L}(f; C_1) \leq \mathcal{L}(f; C_1 \vee C_2) \leq \mathcal{U}(f; C_1 \vee C_2) \leq \mathcal{U}(f; C_2). \quad \blacksquare$$

The Lemma I.1.5 really depends only on properties of our order relation and not on the properties of vol(I). In contrast, the next lemma depends on the continuity of volume with respect to side-lengths of rectangles.

I.1.7 Lemma. *Let C be a non-overlapping, finite, exact cover of the rectangle J and $f : J \longrightarrow \mathbb{R}$ a bounded function. Then for each $\epsilon > 0$ there is a $\delta > 0$ such that*

$$\mathcal{U}(f; C') \leq \mathcal{U}(f; C) + \epsilon \quad \text{and} \quad \mathcal{L}(f; C') \geq \mathcal{L}(f; C) - \epsilon$$

whenever C' is a non-overlapping finite exact cover of J with the property that $\|C'\| < \delta$.

PROOF: In view of (I.1.6), we need only consider the RIEMANN upper sums.

Let $J = \prod_1^N [c_k, d_k]$. Given a $\delta > 0$ and a rectangle $I = \prod_1^N [a_k, b_k]$, define $I_k^-(\delta)$ and $I_k^+(\delta)$ to be the rectangles

$$[c_1, d_1] \times \cdots \times [a_k - \delta, a_k + \delta] \times \cdots \times [c_N, d_N]$$

and

$$[c_1, d_1] \times \cdots \times [b_k - \delta, b_k + \delta] \times \cdots \times [c_N, d_N],$$

respectively. Then for any rectangle $I' \subseteq J$ with $\operatorname{diam}(I') < \delta$, either $I' \subseteq I$ for some $I \in \mathcal{C}$ or

$$I' \subseteq I_k^+(\delta) \cup I_k^-(\delta)$$

for some $I \in \mathcal{C}$ and $1 \leq k \leq N$. Now let $\mathcal{C}'$ with $\|\mathcal{C}'\| < \delta$ be given, define

$$\mathcal{A} = \left\{ I' \in \mathcal{C}' : I' \subseteq I \text{ for some } I \in \mathcal{C} \right\}$$

and set $\mathcal{B} = \mathcal{C} \setminus \mathcal{A}$. Then, by Lemma I.1.1,

$$\sum_{I' \in \mathcal{B}} \operatorname{vol}(I') \leq \sum_{I \in \mathcal{C}} \sum_{k=1}^{N} \left(\operatorname{vol}(I_k^+(\delta)) + \operatorname{vol}(I_k^-(\delta)) \right);$$

from which it is clear that there is a $K < \infty$, depending only on J and $\mathcal{C}$, such that

$$\sum_{I' \in \mathcal{B}} \operatorname{vol}(I') \leq K\delta.$$

Hence,

$$\mathcal{U}(f; \mathcal{C}') - \mathcal{U}(f; \mathcal{C}) \leq \mathcal{U}(f; \mathcal{C}') - \mathcal{U}(f; \mathcal{C} \vee \mathcal{C}')$$

$$\leq \sum_{I' \in \mathcal{B}} \left[\left(\sup_{x \in I'} f(x) \right) \operatorname{vol}(I') - \sum_{I \in \mathcal{C}} \left(\sup_{x \in I \cap I'} f(x) \right) \operatorname{vol}(I \cap I') \right]$$

$$\leq 2\|f\|_u \sum_{I' \in \mathcal{B}} \operatorname{vol}(I') \leq 2K\|f\|_u \delta.$$

(In the preceding, $\operatorname{vol}(R) \equiv 0$ if R is either the empty set or a rectangle with one or more sides of length 0. Also, we have introduced the notation, to be used throughout, that $\|f\|_u$ denotes the **uniform norm of** f: the supremum of f over the set on which f is defined.) From the preceding it is clear how to choose δ for any given $\epsilon > 0$. ∎

As an essentially immediate consequence of Lemma I.1.7, we have the following theorem.

I.1.8 Theorem. *Let* $f : J \longrightarrow \mathbb{R}$ *be a bounded function on the rectangle* J. *Then*

$$\lim_{\|C\|\to 0} \mathcal{L}(f;C) = \sup_{C} \mathcal{L}(f;C) \quad \text{and} \quad \lim_{\|C\|\to 0} \mathcal{U}(f;C) = \inf_{C} \mathcal{U}(f;C),$$

where C *runs over non-overlapping, finite, exact covers of* J. *In particular, at least for bounded* f *'s,* (I.1.4) *is a necessary and sufficient condition for* RIEMANN *integrability. Moreover, if* (I.1.4), *then*

$$(R) \int_{J} f(x) \, dx = \sup_{C} \mathcal{L}(f;C) = \inf_{C} \mathcal{U}(f;C).$$

I.1.9 Exercise.

Prove Theorem I.1.8. Next, suppose that f and g are RIEMANN integrable functions on J. Show that $f \vee g \equiv \max\{f,g\}$, $f \wedge g \equiv \min\{f,g\}$, and, for any α, $\beta \in \mathbb{R}$, $\alpha f + \beta g$ are all RIEMANN integrable on J and that

$$(R) \int_{J} (f \vee g)(x) \, dx \geq \left((R) \int_{J} f(x) \, dx \right) \vee \left((R) \int_{J} g(x) \, dx \right),$$

$$(R) \int_{J} (f \wedge g)(x) \, dx \leq \left((R) \int_{J} f(x) \, dx \right) \wedge \left((R) \int_{J} g(x) \, dx \right),$$

and

$$(R) \int_{J} (\alpha f + \beta g)(x) \, dx = \alpha \, (R) \int_{J} f(x) \, dx + \beta \, (R) \int_{J} g(x) \, dx.$$

Conclude, in particular, that if f and g are RIEMANN integrable on J and $f \leq g$, then $(R) \int_{J} f(x) \, dx \leq (R) \int_{J} g(x) \, dx$.

I.1.10 Exercise.

Show that if f is a bounded real-valued function on the rectangle J, then f is RIEMANN integrable if and only if for each $\epsilon > 0$ there is a $\delta > 0$ such that

(I.1.11) $$\sum_{\{I \in C : \sup_I f - \inf_I f > \epsilon\}} \text{vol}(I) < \epsilon$$

whenever $\|C\| < \delta$. (We use $\sup_I f$ and $\inf_I f$ to denote $\sup_{x \in I} f(x)$ and $\inf_{x \in I} f(x)$, respectively.) As a consequence, show that a bounded f on J is RIEMANN integrable if it is continuous on J at all but a finite number of points. (See Section IV.1 for more information on this subject.)

I.1.12 Exercise.

Show that the condition in Exercise I.1.10 can be replaced by the condition that for each $\epsilon > 0$ there exists a $\mathcal{C}$ for which (I.1.11) holds.

I.2. Riemann-Stieltjes Integration.

In Section I.1, we developed the classical integration theory with respect to the standard notion of EUCLIDean volume. In the present section, we will extend the classical theory, at least for integrals in one dimension, to cover more general notions of volume.

Let $J = [a, b]$ be a interval in $\mathbb{R}$ and let ϕ and ψ be real-valued functions on J. Given a non-overlapping, finite, exact cover $\mathcal{C}$ of J by closed intervals I and a $\xi \in \Xi(\mathcal{C})$, define the **Riemann sum of ϕ over $\mathcal{C}$ with respect to ψ relative to ξ** to be

$$\mathcal{R}(\phi|\psi; \mathcal{C}, \xi) = \sum_{I \in \mathcal{C}} \phi(\xi(I)) \Delta_I \psi,$$

where $\Delta_I \psi \equiv \psi(I^+) - \psi(I^-)$ and I^+ and I^- denote, respectively, the right and left hand end-points of the interval I. Obviously, when $\psi(x) = x, x \in J$, $\mathcal{R}(\phi|\psi; \mathcal{C}, \xi) = \mathcal{R}(\phi; \mathcal{C}, \xi)$. Thus, it is consistent for us to say that ϕ is **Riemann integrable on J with respect to ψ**, or, more simply, **ψ-Riemann integrable on J**, if there is a number A with the property that for each $\epsilon > 0$ there is a $\delta > 0$ such that

(I.2.1) $$\sup_{\xi \in \Xi(\mathcal{C})} |\mathcal{R}(\phi|\psi; \mathcal{C}, \xi) - A| < \epsilon$$

whenever $\mathcal{C}$ is a non-overlapping, finite, exact cover of J satisfying $\|\mathcal{C}\| < \delta$. Assuming that ϕ is ψ-RIEMANN integrable on J, we will call the number A in (I.2.1) the **Riemann–Stieltjes integral of ϕ on J with respect to ψ** and will use

$$(R) \int_J \phi(x) \, d\psi(x)$$

to denote A.

I.2.2 Examples.

i) If $\phi \in C(J)$ and $\psi \in C^1(J)$ (i.e., ψ is continuously differentiable on J), then one can use the Mean Value Theorem to check that ϕ is ψ-RIEMANN integrable on J and that

(I.2.3) $$(R) \int_J \phi(x) \, d\psi(x) = (R) \int_J \phi(x) \psi'(x) \, dx.$$

ii) If there exist $a = a_0 < a_1 < \cdots < a_n = b$ such that ψ is constant on each of the intervals (a_{m-1}, a_m), then every $\phi \in C([a, b])$ is ψ-RIEMANN integrable on $[a, b]$, and

(I.2.4) $$(R) \int_{[a,b]} \phi(x) \, d\psi(x) = \sum_{m=0}^{n} \phi(a_m) d_m,$$

where $d_0 = \psi(a+) - \psi(a)$, $d_m = \psi(a_m+) - \psi(a_m-)$ for $1 \le m \le n-1$, and $d_n = \psi(b) - \psi(b-)$. (We use $f(x+)$ and $f(x-)$ to denote the right and left limits of f at x.)

iii) If both $(R) \int_J \phi_1(x) \, d\psi(x)$ and $(R) \int_J \phi_2(x) \, d\psi(x)$ exist (i.e., ϕ_1 and ϕ_2 are both ψ-RIEMANN integrable on J), then for all real numbers α and β, $(\alpha\phi_1 + \beta\phi_2)$ is ψ-RIEMANN integrable on J and

(I.2.5)
$$(R) \int_J (\alpha\phi_1 + \beta\phi_2)(x) \, d\psi(x)$$
$$= \alpha \left((R) \int_J \phi_1(x) \, d\psi(x) \right) + \beta \left((R) \int_J \phi_2(x) \, d\psi(x) \right).$$

iv) If $J = J_1 \cup J_2$ where $J_1^\circ \cap J_2^\circ = \emptyset$ and if ϕ is ψ-RIEMANN integrable on J, then both $(R) \int_{J_1} \phi(x) \, d\psi(x)$ and $(R) \int_{J_2} \phi(x) \, d\psi(x)$ exist and

(I.2.6) $$(R) \int_J \phi(x) \, d\psi(x) = (R) \int_{J_1} \phi(x) \, d\psi(x) + (R) \int_{J_2} \phi(x) \, d\psi(x).$$

All the assertions made in Examples I.2.2 are reasonably straightforward consequences of the definition of RIEMANN integrability. Not so obvious, but terribly important, is the following theorem which shows that the notion of RIEMANN integrability is *symmetric* in ϕ and ψ.

I.2.7 Theorem. (INTEGRATION BY PARTS) *If ϕ is ψ-RIEMANN integrable on $J = [a, b]$, then ψ is ϕ-RIEMANN integrable on J and*

(I.2.8) $$(R) \int_J \psi(x) \, d\phi(x) = \psi(b)\phi(b) - \psi(a)\phi(a) - (R) \int_J \phi(x) \, d\psi(x).$$

PROOF: Let $\mathcal{C} = \{[\alpha_{m-1}, \alpha_m] : 1 \leq m \leq n\}$, where $a = \alpha_0 < \cdots < \alpha_n = b$; and let $\xi \in \Xi(\mathcal{C})$ with $\xi([\alpha_{m-1}, \alpha_m]) = \beta_m \in [\alpha_{m-1}, \alpha_m]$. Set $\beta_0 = a$ and $\beta_{n+1} = b$. Then

$$\mathcal{R}(\psi|\phi; \mathcal{C}, \xi) = \sum_{m=1}^{n} \psi(\beta_m)\big(\phi(\alpha_m) - \phi(\alpha_{m-1})\big)$$

$$= \sum_{m=1}^{n} \psi(\beta_m)\phi(\alpha_m) - \sum_{m=0}^{n-1} \psi(\beta_{m+1})\phi(\alpha_m)$$

$$= \psi(\beta_n)\phi(\alpha_n) - \sum_{m=1}^{n-1} \phi(\alpha_m)\big(\psi(\beta_{m+1}) - \psi(\beta_m)\big) - \psi(\beta_1)\phi(\alpha_0)$$

$$= \psi(b)\phi(b) - \psi(a)\phi(a) - \sum_{m=0}^{n} \phi(\alpha_m)\big(\psi(\beta_{m+1}) - \psi(\beta_m)\big)$$

$$= \psi(b)\phi(b) - \psi(a)\phi(a) - \mathcal{R}(\phi|\psi; \mathcal{C}', \xi'),$$

where $\mathcal{C}' = \{[\beta_{m-1}, \beta_m] : 1 \leq m \leq n+1 \text{ and } \beta_{m-1} < \beta_m\}$ and $\xi' \in \Xi(\mathcal{C}')$ is defined by $\xi'([\beta_m, \beta_{m+1}]) = \alpha_m$ when $\beta_m < \beta_{m+1}$. Noting that $\|\mathcal{C}'\| \leq 2\|\mathcal{C}\|$, one now sees how to get the desired conclusion. ∎

Although the preceding theorem indicates that it is natural to consider ϕ and ψ as playing symmetric roles in the theory of RIEMANN–STIELTJES integration, it turns out that, in practice, one wants to impose a condition on ψ which will guarantee that every $\phi \in C(J)$ is RIEMANN integrable with respect to ψ and that, in addition,

$$(I.2.9) \qquad \left| (R)\int_J \phi(x)\, d\psi(x) \right| \leq K_\psi \|\phi\|_u$$

for some $K_\psi < \infty$ and all ϕ which are ψ-RIEMANN integrable on J. Example i) in Examples I.2.2 tells us that one condition on ψ which guarantees the ψ-RIEMANN integrability of every continuous ϕ is that $\psi \in C^1(J)$. Moreover, from (I.2.3), it is an easy matter to check that in this case (I.2.9) holds with $K_\psi = \|\psi'\|_u(b - a)$. On the other hand, example ii) makes it clear that ψ need not be even continuous, much less differentiable, in order that RIEMANN integration with respect to ψ have the above properties. The following result emphasizes this same point.

I.2.10 Theorem. *Let ψ be non-decreasing on J. Then every $\phi \in C(J)$ is ψ-RIEMANN integrable on J. In addition, if ϕ is non-negative and ψ-RIEMANN* *integrable on* J, *then* $(R) \int_J \phi(x)\,d\psi(x) \geq 0$. *In particular,* (I.2.9) *holds with* $K_\psi = \Delta_J \psi$.

PROOF: The fact that $(R) \int_J \phi(x)\,d\psi(x) \geq 0$ if ϕ is a non-negative function which is ψ-RIEMANN integrable on J follows immediately from the fact the $\mathcal{R}(\phi|\psi;\mathcal{C},\xi) \geq 0$ for any $\mathcal{C}$ and $\xi \in \Xi(\mathcal{C})$. Applying this to the function $\|\phi\|_{\mathrm{u}} - \phi$ and using the linearity property in iii) of Example I.2.2, we conclude that (I.2.9) holds with $K_\psi = \Delta_J \psi$. Thus, all that we have to do is check that every $\phi \in C(J)$ is ψ-RIEMANN integrable on J.

Let $\phi \in C(J)$ be given and define

$$\mathcal{U}(\phi|\psi;\mathcal{C}) = \sum_{I \in \mathcal{C}} (\sup_I \phi)\Delta_I \psi \quad \text{and} \quad \mathcal{L}(\phi|\psi;\mathcal{C}) = \sum_{I \in \mathcal{C}} (\inf_I \phi)\Delta_I \psi$$

for $\mathcal{C}$ and $\xi \in \Xi(\mathcal{C})$. Then, just as in Section I.1,

$$\mathcal{L}(\phi|\psi;\mathcal{C}) \leq \mathcal{R}(\phi|\psi;\mathcal{C},\xi) \leq \mathcal{U}(\phi|\psi;\mathcal{C})$$

for any $\xi \in \Xi(\mathcal{C})$. In addition (cf. Lemma I.1.5), for any pair $\mathcal{C}_1$ and $\mathcal{C}_2$ one has that $\mathcal{L}(\phi|\psi;\mathcal{C}_1) \leq \mathcal{U}(\phi|\psi;\mathcal{C}_2)$. Finally, for any $\mathcal{C}$:

$$\mathcal{U}(\phi|\psi;\mathcal{C}) - \mathcal{L}(\phi|\psi;\mathcal{C}) \leq \omega(\|\mathcal{C}\|)\Delta_J \psi,$$

where

$$\omega(\delta) \equiv \sup \big\{ |\phi(y) - \phi(x)| : x, y \in J \text{ and } |y - x| \leq \delta \big\}$$

is the **modulus of continuity** of ϕ. Hence

$$\lim_{\|\mathcal{C}\| \to 0} \big(\mathcal{U}(\phi|\psi;\mathcal{C}) - \mathcal{L}(\phi|\psi;\mathcal{C})\big) = 0.$$

But this means that for every $\epsilon > 0$ there is a $\delta > 0$ such that

$$\mathcal{U}(\phi|\psi;\mathcal{C}) - \mathcal{U}(\phi|\psi;\mathcal{C}') \leq \mathcal{U}(\phi|\psi;\mathcal{C}) - \mathcal{L}(\phi|\psi;\mathcal{C}) < \epsilon$$

no matter what $\mathcal{C}'$ is chosen so long as $\|\mathcal{C}\| < \delta$. From these it is clear that both

$$\lim_{\|\mathcal{C}\| \to 0} \mathcal{U}(\phi|\psi;\mathcal{C}) \quad \text{and} \quad \lim_{\|\mathcal{C}\| \to 0} \mathcal{L}(\phi|\psi;\mathcal{C})$$

exist and are equal. ∎

One obvious way in which to extend the preceding result is to note that if ϕ is RIEMANN integrable on J with respect to both ψ_1 and ψ_2, then it is RIEMANN integrable on J with respect to $\psi_2 - \psi_1$ and

$$(R) \int_J \phi(x) \, d(\psi_2 - \psi_1)(x) = (R) \int_J \phi(x) \, d\psi_2(x) - (R) \int_J \phi_1(x) \, d\psi_1(x).$$

(This can be seen directly or as a consequence of Theorem I.2.7 combined with **iii**) in Examples I.2.2.) In particular, we have the following corollary to Theorem I.2.10.

I.2.11 Corollary. *If $\psi = \psi_2 - \psi_1$ where ψ_1 and ψ_2 are non-decreasing functions on J, then every $\phi \in C(J)$ is RIEMANN integrable with respect to ψ and (I.2.9) holds with $K_\psi = \Delta_J \psi_1 + \Delta_J \psi_2$.*

We are now going to embark on a program which will show that, at least among ψ's which are right continuous on J° and have left limits at each point in $J \setminus \{J^-\}$, the ψ's in Corollary I.2.11 are the only ones with the properties that every $\phi \in C(J)$ is ψ-RIEMANN integrable on J and (I.2.9) holds for some $K_\psi < \infty$. The first step is to provide an alternative description of those ψ's which can be expressed as the difference of two non-decreasing functions. To this end, let ψ be a real-valued function on J and define

$$S(\psi; \mathcal{C}) = \sum_{I \in \mathcal{C}} |\Delta_I \psi|$$

for any non-overlapping, finite, exact cover $\mathcal{C}$ of J. Clearly:

$$S(\alpha \psi; \mathcal{C}) = |\alpha| S(\psi; \mathcal{C}) \quad \text{for all} \quad \alpha \in \mathbb{R},$$

$$S(\psi_1 + \psi_2; \mathcal{C}) \leq S(\psi_1; \mathcal{C}) + S(\psi_2; \mathcal{C}) \quad \text{for all} \quad \psi_1 \text{ and } \psi_2,$$

and

$$S(\psi; \mathcal{C}) = |\Delta_J \psi|$$

if ψ is monotone on J. Moreover, if $\mathcal{C}$ is given and $\mathcal{C}'$ is obtained from $\mathcal{C}$ by replacing one of the I's in $\mathcal{C}$ by a pair $\{I_1, I_2\}$, where $I = I_1 \cup I_2$ and $I_1^\circ \cap I_2^\circ = \emptyset$, then, by the triangle inequality,

$$S(\psi; \mathcal{C}') - S(\psi; \mathcal{C})$$
$$= |\psi(I_1^+) - \psi(I_1^-)| + |\psi(I_2^+) - \psi(I_2^-)| - |\psi(I^+) - \psi(I^-)| \geq 0.$$

Hence we see that

$$S(\psi; C) \leq S(\psi; C') \quad \text{for} \quad C \leq C'.$$

We now define the **variation of** ψ **on** J to be the number (possibly infinite)

(I.2.12)
$$\text{Var}(\psi; J) \equiv \sup_C S(\psi; C),$$

where the C's run over all non-overlapping, finite, exact covers of J. Also, we say that ψ has **bounded variation on** J if $\text{Var}(\psi; J) < \infty$. It should be clear that if $\psi = \psi_2 - \psi_1$ for non-decreasing ψ_1 and ψ_2 on J, then ψ has bounded variation on J and $\text{Var}(\psi; J) \leq \Delta_J \psi_1 + \Delta_J \psi_2$. What is less obvious is that every ψ having bounded variation on J can be expressed as the difference of two non-decreasing functions. In order to prove this fact, we introduce

$$S_+(\psi; C) = \sum_{I \in C} (\Delta_I \psi)^+$$

and

$$S_-(\psi; C) = \sum_{I \in C} (\Delta_I \psi)^-,$$

where $\alpha^+ \equiv \alpha \vee 0$ and $\alpha^- \equiv -(\alpha \wedge 0)$ for $\alpha \in \mathbb{R}$. Also, we call the numbers

$$\text{Var}_+(\psi; J) \equiv \sup_C S_+(\psi; C) \quad \text{and} \quad \text{Var}_-(\psi; J) \equiv \sup_C S_-(\psi; C)$$

the **positive variation** and the **negative variation** of ψ on J. Noting that

(I.2.13)
$$2S_\pm(\psi; C) = S(\psi; C) \pm \Delta_J \psi$$
$$S_+(\psi; C) - S_-(\psi; C) = \Delta_J \psi$$
$$S_+(\psi; C) + S_-(\psi; C) = S(\psi; C)$$

for any C, we see that

$$S_\pm(\psi; C) \leq S_\pm(\psi; C'), \qquad C \leq C',$$

and that

$$\text{Var}_+(\psi; J) < \infty \iff \text{Var}(\psi; J) < \infty \iff \text{Var}_-(\psi; J) < \infty.$$

I.2.14 Lemma. *If* $\mathrm{Var}(\psi; J) < \infty$, *then*

(I.2.15) $$\mathrm{Var}_+(\psi; J) + \mathrm{Var}_-(\psi; J) = \mathrm{Var}(\psi; J)$$

and

(I.2.16) $$\mathrm{Var}_+(\psi; J) - \mathrm{Var}_-(\psi; J) = \Delta_J \psi$$

PROOF: By the middle relation in (I.2.13), we see that

$$\mathcal{S}_\pm(\psi; \mathcal{C}) \le \mathrm{Var}_\mp(\psi; J) \pm \Delta_J \psi.$$

Hence,

$$\mathrm{Var}_\pm(\psi; J) \le \mathrm{Var}_\mp(\psi; J) \pm \Delta_J \psi;$$

and so (I.2.16) has been proved. Moreover, (I.2.16) combined with the middle relation in (I.2.13) leads to

$$\mathrm{Var}_+(\psi; J) - \mathcal{S}_+(\psi; \mathcal{C}) = \mathrm{Var}_-(\psi; J) - \mathcal{S}_-(\psi; \mathcal{C})$$

for any $\mathcal{C}$. In particular, there is a sequence $\{\mathcal{C}_n\}_1^\infty$ such that $\mathcal{S}_+(\psi; \mathcal{C}_n) \longrightarrow$ $\mathrm{Var}_+(\psi; J)$ as $n \to \infty$ and, at the same time, $\mathcal{S}_-(\psi; \mathcal{C}_n) \longrightarrow \mathrm{Var}_-(\psi; J)$. Hence, by the last relation in (I.2.13), we see that

$$\mathrm{Var}_+(\psi; J) + \mathrm{Var}_-(\psi; J) \le \varlimsup_{n \to \infty} \mathcal{S}(\psi; \mathcal{C}_n) \le \mathrm{Var}(\psi; J).$$

At the same time, by that same relation in (I.2.13),

$$\mathcal{S}(\psi; \mathcal{C}) = \mathcal{S}_+(\psi; \mathcal{C}) + \mathcal{S}_-(\psi; \mathcal{C}) \le \mathrm{Var}_+(\psi; J) + \mathrm{Var}_-(\psi; J)$$

for every $\mathcal{C}$. When combined with the preceding, this completes the proof of (I.2.15). ∎

I.2.17 Lemma. *If* ψ *has bounded variation on* $[a, b]$ *and* $a < c < b$, *then*

$$\mathrm{Var}_\pm(\psi; [a, b]) = \mathrm{Var}_\pm(\psi; [a, c]) + \mathrm{Var}_\pm(\psi; [c, b]),$$

and therefore also $\mathrm{Var}(\psi; [a, b]) = \mathrm{Var}(\psi; [a, c]) + \mathrm{Var}(\psi; [c, b])$.

PROOF: Because of the (I.2.15) and (I.2.16), we see that it suffices to check the equality only for "Var" itself. But if $\mathcal{C}_1$ and $\mathcal{C}_2$ are non-overlapping, finite, exact covers of $[a, c]$ and $[c, b]$, then $\mathcal{C} = \mathcal{C}_1 \cup \mathcal{C}_2$ is a non-overlapping, finite, exact cover of $[a, b]$; and so

$$\mathcal{S}(\psi; \mathcal{C}_1) + \mathcal{S}(\psi; \mathcal{C}_2) = \mathcal{S}(\psi; \mathcal{C}) \le \mathrm{Var}(\psi; [a, b]).$$

Hence $\mathrm{Var}(\psi; [a, c]) + \mathrm{Var}(\psi; [c, b]) \le \mathrm{Var}(\psi; [a, b])$. On the other hand, if $\mathcal{C}$ is a non-overlapping, finite, exact cover of $[a, b]$, then it is easy to construct non-overlapping, finite, exact covers $\mathcal{C}_1$ and $\mathcal{C}_2$ of $[a, c]$ and $[c, b]$ such that $\mathcal{C} \le \mathcal{C}_1 \cup \mathcal{C}_2$. Hence,

$$\mathcal{S}(\psi; \mathcal{C}) \le \mathcal{S}(\psi; \mathcal{C}_1 \cup \mathcal{C}_2) = \mathcal{S}(\psi; \mathcal{C}_1) + \mathcal{S}(\psi; \mathcal{C}_2) \le \mathrm{Var}(\psi; [a, c]) + \mathrm{Var}(\psi; [c, b]).$$

Since this is true for every $\mathcal{C}$, the asserted equality is now proved. ∎

We have now proved the following decomposition theorem for functions having bounded variation.

I.2.18 Theorem. *Let $\psi : J \longrightarrow \mathbb{R}$ be given. Then ψ has bounded variation on J if and only if there exist non-decreasing functions ψ_1 and ψ_2 on J such that $\psi = \psi_2 - \psi_1$. In fact, if ψ has bounded variation on $J = [a, b]$ and we define $\psi_{\pm}(x) = \mathrm{Var}_{\pm}(\psi; [a, x])$ for $x \in J$, then ψ_+ and ψ_- are non-decreasing and $\psi(x) = \psi(a) + \psi_+(x) - \psi_-(x)$, $x \in J$. Finally, if ψ has bounded variation on J, then every $\phi \in C(J)$ is* RIEMANN *integrable on J with respect to ψ and*

$$(I.2.19) \qquad \left| (R) \int_J \phi(x)\, d\psi(x) \right| \leq \mathrm{Var}(\psi; J) \|\phi\|_{\mathrm{u}}.$$

In order to complete our program, we need one more elementary fact.

I.2.20 Lemma. *If $\psi : J \longrightarrow \mathbb{R}$ has a right limits in $\mathbb{R}$ at every $x \in J \setminus \{J^+\}$ and a left limit in $\mathbb{R}$ at every $x \in J \setminus \{J^-\}$, then ψ is bounded and*

$$\mathrm{card}\left(\left\{ x \in J^\circ : |\psi(x) - \psi(x+)| \vee |\psi(x) - \psi(x-)| \geq \epsilon \right\} \right) < \infty \quad \text{for each} \quad \epsilon > 0.$$

In particular, ψ has at most countably many discontinuities. Also, if $\tilde{\psi}(x) \equiv \psi(x+)$ for $x \in J^\circ$ and $\tilde{\psi}(x) = \psi(x)$ for $x \in \{J^-, J^+\}$, then $\tilde{\psi}$ is right-continuous on J°, has a left limit in $\mathbb{R}$ at every $x \in J \setminus \{J^-\}$, and coincides with ψ at all points where ψ is continuous. Thus, if ϕ is RIEMANN *integrable on J with respect to both ψ and $\tilde{\psi}$, then $(R) \int_J \phi(x)\, d\tilde{\psi}(x) = (R) \int_J \phi(x)\, d\psi(x)$. Finally, if $\phi \in C(J)$ is* RIEMANN *integrable on J with respect to ψ, then it is also* RIEMANN *integrable on J with respect to $\tilde{\psi}$.*

PROOF: Suppose that ψ were unbounded. Then we could find $\{x_n\}_1^\infty \subseteq J$ so that $|\psi(x_n)| \longrightarrow \infty$ as $n \to \infty$; and clearly there is no loss in generality if we assume that $x_{n+1} < x_n$ for all $n \geq 1$. But this would mean that $|\psi(x+)| = \infty$, where $x = \lim_{n \to \infty} x_n$, and so no such sequence can exist. Thus ψ must be bounded. The proof that $\mathrm{card}\{x \in J : |\psi(x) - \psi(x+)| \vee |\psi(x) - \psi(x-)| \geq \epsilon\} < \infty$ is similar. Namely, if not, then we could assume that there exists a strictly decreasing sequence $\{x_n\} \subseteq J^\circ$ with limit $x \in J$ such that $|\psi(x_n) - \psi(x_n+)| \vee |\psi(x_n) - \psi(x_n-)| \geq \epsilon$ for each $n \geq 1$. But then, for each $n \geq 1$, we could find $x'_n \in (x, x_n)$ and $x''_n \in (x_n, x_n + \frac{1}{n}) \cap J^\circ$ so that $|\psi(x_n) - \psi(x'_n)| \vee |\psi(x_n) - \psi(x''_n)| \geq \epsilon/2$; and clearly this contradicts the existence in $\mathbb{R}$ of $\psi(x+)$.

The preceding makes it obvious that ψ can be discontinuous at only countably many points. In addition it is clear that $\tilde{\psi}(x\pm) = \psi(x\pm)$ for all $x \in J^\circ$. To prove the equality of RIEMANN integrals with respect to ψ and $\tilde{\psi}$ of ϕ's which are RIEMANN integrable with respect to both, note that, because ψ coincides with $\tilde{\psi}$ on $\{J^-, J^+\}$ as well as on a dense subset of J°, we can always evaluate these integrals using RIEMANN sums which are the same whether they are computed with respect to ψ or to $\tilde{\psi}$.

Finally, we must show that if $\phi \in C(J)$ is RIEMANN integrable with respect to ψ, then it also is with respect to $\tilde{\psi}$. To do this, it clearly suffices to show that for any $C, \xi \in \Xi(C)$ and $\epsilon > 0$, there is a C' and a $\xi' \in \Xi(C')$ such that $\|C'\| \leq 2\|C\|$ and $|\mathcal{R}(\phi|\tilde{\psi}; C, \xi) - \mathcal{R}(\phi|\psi; C', \xi')| < \epsilon$. To this end, let $J^- = c_0 < \cdots < c_{n+1} = J^+$ be chosen so that $C = \{[c_k, c_{k+1}] : 0 \leq k \leq n\}$. For $0 < \alpha < c_{n+1} - c_n$, set

$$c_{k,\alpha} = \begin{cases} c_k & \text{if} \quad k \in \{0, n+1\} \\ c_k + \alpha & \text{if} \quad 1 \leq k \leq n. \end{cases}$$

and let $C_\alpha = \{[c_{k,\alpha}, c_{k+1,\alpha}] : 0 \leq k \leq n\}$. Clearly $\|C_\alpha\| \leq 2\|C\|$. Next, define $\xi_\alpha([c_{k,\alpha}, c_{k+1,\alpha}]) = c_{k,\alpha}$ for $0 \leq k \leq n$. Then, because ϕ is continuous and $\tilde{\psi}$ is right-continuous on J°,

$$\mathcal{R}(\phi|\tilde{\psi}; C, \xi) = \lim_{\alpha \searrow 0} \mathcal{R}(\phi|\tilde{\psi}; C_\alpha, \xi_\alpha).$$

At the same time,

$$\mathcal{R}(\phi|\tilde{\psi}; C_\alpha, \xi_\alpha) = \mathcal{R}(\phi|\psi; C_\alpha, \xi_\alpha)$$

for all but a countable number of α's. Thus, for any $\epsilon > 0$, there are arbitrarily small α's for which $|\mathcal{R}(\phi|\psi; C, \xi) - \mathcal{R}(\phi|\tilde{\psi}; C_\alpha, \xi_\alpha)| < \epsilon$. ∎

I.2.21 Remark.

Any ψ having bounded variation on J certainly satisfies the hypotheses of Lemma I.2.20. Moreover, if $\tilde{\psi}$ is defined accordingly, then $\text{Var}(\tilde{\psi}; J) \leq \text{Var}(\psi; J)$.

I.2.22 Theorem. *Let ψ be a function on J which satisfies the hypotheses of Lemma I.2.20, and define $\tilde{\psi}$ accordingly. If every $\phi \in C(J)$ is RIEMANN integrable on J with respect to ψ and if there is a $K < \infty$ such that*

$$(\text{I.2.23}) \qquad \left| (R) \int_J \phi(x) \, d\psi(x) \right| \leq K \|\phi\|_{\mathrm{u}}, \qquad \phi \in C(J),$$

then $\tilde{\psi}$ has bounded variation on J and

$$\text{Var}(\tilde{\psi}; J) = \sup \left\{ (R) \int_J \phi(x) \, d\psi(x) : \phi \in C(J) \text{ and } \|\phi\|_{\mathrm{u}} = 1 \right\}$$

$$(\text{I.2.24}) \qquad = \sup \left\{ (R) \int_J \phi(x) \, d\tilde{\psi}(x) : \phi \in C(J) \text{ and } \|\phi\|_{\mathrm{u}} = 1 \right\}.$$

In particular, if ψ itself is right-continuous on $J°$, then ψ has bounded variation on J if and only if every $\phi \in C(J)$ is RIEMANN *integrable on J with respect to ψ and (I.2.23) holds for some $K < \infty$; in which case $\mathrm{Var}(\psi; J)$ is the optimal choice of K.*

PROOF: In view of what we already know, all that we have to do is check that for each $\mathcal{C}$ and $\epsilon > 0$ there is a $\phi \in C(J)$ such that $\|\phi\|_u = 1$ and $\mathcal{S}(\tilde{\psi}; \mathcal{C}) \le (R) \int_J \phi(x) \, d\psi(x) + \epsilon$. Moreover, because $\tilde{\psi}$ is right continuous, we may and will assume that $\mathcal{C} = \{[c_k, c_{k+1}] : 0 \le k \le n\}$ where $J^- = c_0 < \cdots < c_{n+1} = J^+$ and c_k is a point of continuity of ψ for each $1 \le k \le n$.

Given $0 < \alpha < \min_{1 \le k \le n}(c_{k+1} - c_k)/2$, define $\phi_\alpha \in C(J)$ so that: $\phi_\alpha(x) = \mathrm{sgn}(\Delta_{[c_0, c_1]}\psi)$ for $x \in [c_0, c_1 - \alpha]$, $\phi_\alpha(x) = \mathrm{sgn}(\Delta_{[c_k, c_{k+1}]}\psi)$ for $x \in [c_k + \alpha, c_{k+1} - \alpha]$ and $1 \le k < n$, $\phi_\alpha(x) = \mathrm{sgn}(\Delta_{[c_n, c_{n+1}]}\psi)$ for $x \in [c_n + \alpha, c_{n+1}]$, and ϕ_α is linear on each of the intervals $[c_k - \alpha, c_k + \alpha]$, $1 \le k \le n$. (The **signum function** $t \in \mathbb{R} \longmapsto \mathrm{sgn}(t)$ is defined so that $\mathrm{sgn}(t)$ is -1 or 1 according to whether $t < 0$ or $t \ge 0$.) Then, by **iv)** in Examples I.2.2,

$$(R) \int_J \phi_\alpha(x) \, d\psi(x) - \mathcal{S}(\tilde{\psi}; \mathcal{C})$$

$$= \sum_{k=0}^n (R) \int_{[c_k, c_{k+1}]} (\phi_\alpha(x) - \mathrm{sgn}(\Delta_{[c_k, c_{k+1}]}\psi)) \, d\psi(x)$$

$$= \sum_{k=1}^n \left[(R) \int_{[c_k - \alpha, c_k]} (\phi_\alpha(x) - \phi_\alpha(c_k - \alpha)) \, d\psi(x) \right.$$

$$\left. + (R) \int_{[c_k, c_k + \alpha]} (\phi_\alpha(x) - \phi_\alpha(c_k + \alpha)) \, d\psi(x) \right].$$

For each $1 \le k \le n$, either $\phi_\alpha \equiv \phi_\alpha(c_k - \alpha)$ on $[c_k - \alpha, c_k + \alpha]$, in which case the corresponding term does not contribute to the preceding sum, or $\phi_\alpha(c_k) = 0$ and $\phi'_\alpha \equiv (\phi_\alpha(c_k + \alpha) - \phi_\alpha(c_k - \alpha))/2\alpha$ on $[c_k - \alpha, c_k + \alpha]$. In the latter case, we apply Theorem I.2.7 and equation (I.2.3) to show that

$$(R) \int_{[c_k - \alpha, c_k]} (\phi_\alpha(x) - \phi_\alpha(c_k - \alpha)) \, d\psi(x)$$

$$+ (R) \int_{[c_k, c_k + \alpha]} (\phi_\alpha(x) - \phi_\alpha(c_k + \alpha)) \, d\psi(x)$$

$$= [\phi_\alpha(c_k + \alpha) - \phi_\alpha(c_k - \alpha)] \psi(c_k)$$

$$- \frac{\phi_\alpha(c_k + \alpha) - \phi_\alpha(c_k - \alpha)}{2\alpha} \int_{[c_k - \alpha, c_k + \alpha]} \psi(x) \, dx,$$

which, since ψ is continuous at c_k, clearly tends to 0 as $\alpha \searrow 0$. In other words, we now see that

$$\mathcal{S}(\tilde{\psi}; \mathcal{C}) = \lim_{\alpha \searrow 0} (R) \int_J \phi_\alpha(x) \, d\psi(x),$$

which is all that we had to prove. ∎

I.2.25 Exercise.

Check all of the assertions in Examples I.2.2. The only one which presents a challenge is the assertion in **iv**) that ϕ is RIEMANN integrable on both J_1 and J_2 with respect to ψ.

I.2.26 Exercise.

If ψ is non-decreasing on J, show that a bounded function ϕ is RIEMANN integrable on J with respect to ψ if and only if for every $\epsilon > 0$ there is a $\delta > 0$ such that

$$(I.2.27) \qquad \sum_{\{I \in \mathcal{C} : \sup_I \phi - \inf_I \phi \geq \epsilon\}} \Delta_I \psi < \epsilon$$

whenever $\mathcal{C}$ is a non-overlapping, finite, exact cover of J satisfying $\|\mathcal{C}\| < \delta$. Also, show that when, in addition, $\psi \in C(J)$, the preceding can be replaced by the condition that, for each $\epsilon > 0$, (I.2.27) holds for some $\mathcal{C}$. (**Hint:** for the last part, compare the situation here to the one handled in Lemma I.1.7.)

I.2.28 Exercise.

If $\psi \in C(J)$, show that

$$\mathrm{Var}_\pm(\psi; J) = \lim_{\|\mathcal{C}\| \to 0} \mathcal{S}_\pm(\psi; \mathcal{C}) \ (\in [0, \infty])$$

and conclude that $\mathrm{Var}(\psi, J) = \lim_{\|\mathcal{C}\| \to 0} \mathcal{S}(\psi; \mathcal{C})$. Also, show that if $\psi \in C^1(J)$, then

$$\mathrm{Var}_\pm(\psi; J) = (R) \int_J \psi'(x)^\pm \, dx$$

and therefore that $\mathrm{Var}(\psi; J) = (R) \int_J |\psi'(x)| \, dx$.

I.2.29 Exercise.

Let ψ be a function of bounded variation on the interval $J = [c, d]$, and define the non-decreasing functions ψ_+ and ψ_- accordingly as in Theorem I.2.19. Given any other pair of non-decreasing functions ψ_1 and ψ_2 on J satisfying $\psi = \psi_2 - \psi_1$, show that $\psi_2 - \psi_+$ and $\psi_1 - \psi_-$ are both non-decreasing functions. In particular, this means that $\psi_+ - \psi_+(c) \le \psi_2 - \psi_2(c)$ and $\psi_- - \psi_-(c) \le \psi_1 - \psi_1(c)$ whenever ψ_2 and ψ_1 are non-decreasing functions for which $\psi = \psi_2 - \psi_1$.

Using the preceding, show that

$$\psi_\pm(x+) - \psi_\pm(x) = \Big(\psi(x+) - \psi(x)\Big)^\pm, \quad x \in [c, d),$$

and

$$\psi_\pm(x) - \psi_\pm(x-) = \Big(\psi(x) - \psi(x-)\Big)^\pm, \quad x \in (c, d].$$

Conclude, in particular, that the jumps in $x \longmapsto \mathrm{Var}(\psi; [a, x])$, from both the right and left, coincide with the absolute value of the corresponding jumps in ψ. Hence, ψ is continuous if $x \in J \longmapsto \mathrm{Var}(\psi; [c, x])$ is; and if ψ is continuous, then so are ψ_+ and ψ_- and therefore also $\mathrm{var}(\psi; [c, \cdot])$.

I.2.30 Exercise.

Construct an example of a $\psi \in C([0, 1])$ such that $\mathrm{Var}(\psi; [0, 1]) = \infty$. Also, give an example of a ψ having bounded variation on $[0, 1]$ for which

$$\sup\left\{ (R) \int_{[0,1]} \phi(x)\, d\psi(x) : \phi \in C(J) \text{ and } \|\phi\|_u = 1 \right\} < \mathrm{Var}(\psi; J).$$

Chapter II. Lebesgue's Measure

II.0. The Idea.

In this chapter we construct LEBESGUE's measure on $\mathbb{R}^N$, and in the following chapter we will develop his method of integration. In order to avoid getting lost in the details, it will be important to keep in mind what it is that we are attempting to do. For this reason, we will begin with a brief summary of our goals.

The essence of any theory of integration is a *divide and conquer* strategy. That is, given a space E and a family $\mathcal{B}$ of subsets $\Gamma \subseteq E$ for which one has a *reasonable notion* of *measure* assignment $\Gamma \in \mathcal{B} \longmapsto \mu(\Gamma) \in [0, \infty]$, the *integral* of a function $f : E \longrightarrow \mathbb{R}$ should be computed by a prescription containing the following ingredients. In the first place, one has to choose a partition $\mathcal{P}$ of the space E into subsets $\Gamma \in \mathcal{B}$. Secondly, given $\mathcal{P}$, one has to select for each $\Gamma \in \mathcal{P}$ a *typical* value a_Γ of f on Γ. Thirdly, given both the partition $\mathcal{P}$ and the selection

$$\Gamma \in \mathcal{P} \longmapsto a_\Gamma \in \text{Range}(f|_\Gamma),$$

one forms the sum

$$(\text{II.0.1}) \qquad \sum_{\Gamma \in \mathcal{P}} a_\Gamma \, \mu(\Gamma).$$

Finally, by using a limit procedure if necessary, one removes the ambiguity (inherent in the notion of *typical*) by choosing the partitions $\mathcal{P}$ in such a way that the restriction of f to each Γ is increasingly close to a constant.

Obviously, even if we ignore all questions of convergence, the only way in which we can make sense out of (II.0.1) is if we restrict ourselves to either finite or, at worst, countable partitions $\mathcal{P}$. Hence, in general, the final limit procedure will be essential. Be that as it may, when E is itself countable and $\{x\} \in \mathcal{B}$ for every $x \in E$, there is an *obvious* way to avoid the limit step, namely ones chooses $\mathcal{P} = \big\{ \{x\} : x \in E \big\}$ and takes the *integral* to be

$$(\text{II.0.2}) \qquad \sum_{x \in E} f(x) \mu(\{x\}).$$

(We will ignore, for the present, all problems which arise from questions of convergence.) Clearly, this is the idea on which RIEMANN based his theory of integration. On the other hand, RIEMANN's is not the only *obvious* way to proceed, even in the case of countable spaces E. For example, again assuming that E is countable, let $f : E \longrightarrow \mathbb{R}$ be given. Then Range(f) is countable and, assuming that $\Gamma(a) \equiv \{x \in E : f(x) = a\} \in \mathcal{B}$ for every $a \in \mathbb{R}$, LEBESGUE would say that

$$(\text{II.0.3}) \qquad \sum_{a \in \text{Range}(f)} a \, \mu\big(\Gamma(a)\big)$$

is an equally *obvious* candidate for the *integral* of f.

In order to reconcile these two *obvious* definitions, one has to examine the assignment $\Gamma \in \mathcal{B} \longmapsto \mu(\Gamma) \in [0, \infty]$ of *measure*. Indeed, even if E is countable and $\mathcal{B}$ contains every subset of E, (II.0.2) and (II.0.3) give the same answer only if one knows about μ that, for any countable collection $\{\Gamma_n\} \subseteq \mathcal{B}$,

$$(\text{II.0.4}) \qquad \mu\left(\bigcup_n \Gamma_n\right) = \sum_n \mu(\Gamma_n) \quad \text{when} \quad \Gamma_m \cap \Gamma_n = \emptyset \quad \text{for} \quad m \neq n.$$

When E is countable, (II.0.4) is equivalent to taking

$$\mu(\Gamma) = \sum_{x \in \Gamma} \mu(\{x\}), \quad \Gamma \subseteq E.$$

However, when E is uncountable, the property in (II.0.4) becomes highly non-trivial. In fact, it is unquestionably LEBESGUE's most significant achievement to have shown that there are non-trivial *assignments of measure* which enjoy this property.

Having compared LEBESGUE's to RIEMANN's ideas as they appear in the countable setting, we close this introduction to LEBESGUE's theory with a few words about the same comparison for uncountable spaces. For this purpose, we will suppose that $E = [0, 1]$ and, without worrying about exactly which subsets of E are included in $\mathcal{B}$, we will suppose that $\mathcal{B}$ contains not only all open and closed subsets of E but also all the sets which can be obtained, starting from open and closed sets, by countable, set-theoretic operations. Further, we assume that $\Gamma \in \mathcal{B} \longmapsto \mu(\Gamma) \in [0, 1]$ is a mapping which satisfies (II.04). Now let $f : [0, 1] \longrightarrow \mathbb{R}$ be given. In order to integrate f, RIEMANN says that we should divide up $[0, 1]$ into small intervals, choose a representative value of f from each interval, form the associated RIEMANN sum, and then take the limit as the mesh size of the division tends to 0. As we know, his procedure works beautifully so long as the function f respects the topology of the real line; that is, so long as f is sufficiently continuous. However, RIEMANN's procedure is doomed to failure when f does not respect the topology of $\mathbb{R}$. The problem is, of course, that RIEMANN's partitioning procedure is tied to the topology of the reals and is therefore too rigid to accommodate functions which pay little or no attention to that topology. To get around this problem, LEBESGUE tailors his partitioning procedure to the particular function f under consideration. Thus, for a given function f, LEBESGUE might consider the sequence of partitions $\mathcal{P}_n$, $n \in \mathbb{N}$, consisting of the sets

$$\Gamma_{n,k} = \left\{ x \in E : f(x) \in \left[\frac{k}{2^n}, \frac{k+1}{2^n} \right) \right\}, \quad k \in \mathbb{Z}.$$

Obviously, no two values that f takes on any one of the $\Gamma_{n,k}$'s can differ by more than $\frac{1}{2^n}$. Hence, assuming that $\Gamma_{n,k} \in \mathcal{B}$ for every $n \in \mathbb{N}$ and $k \in \mathbb{Z}$ and ignoring convergence problems,

$$\lim_{n \to \infty} \sum_{k \in \mathbb{Z}} \frac{k}{2^n} \mu(\Gamma_{n,k})$$

has just got to be the *integral* of f!

When one hears LEBESGUE's ideas for the first time, one may well wonder what there is left to be done. On the other hand, after a little reflection, some doubts begin to emerge. For example, what is so sacrosanct about LEBESGUE's partitioning suggested in the preceding paragraph and, for instance, why should one have not done the same thing relative to 3^n instead of 2^n? The answer is, of course, that there is nothing to recommend 2^n over 3^n and that it should make no difference which of them is used. Thus, one has to check that it really does not matter, and, once again, the verification entails repeated application of the property in (II.0.4). In fact, it will become increasing evident that LEBESGUE's entire program rests on (II.0.4).

With the preceding comments in mind, it should be clear why we initiate LEBESGUE's program with his proof that there is an interesting μ satisfying (II.0.4) actually exists. To be more precise, the rest of this chapter is devoted to showing that we can define such a μ on a rich class of subsets of $\mathbb{R}^N$ so that $\mu(I) = \text{vol}(I)$ whenever I is a rectangle.

II.1. Existence.

Given a countable (possibly overlapping) cover $\mathcal{C}$ of a subset $\Gamma \subseteq \mathbb{R}^N$ by rectangles I, define $\Sigma(\mathcal{C}) = \sum_{I \in \mathcal{C}} \text{vol}(I) \in [0, \infty]$. We call the number

$$|\Gamma|_e = \inf\Big\{\Sigma(\mathcal{C}) : \Gamma \subseteq \bigcup \mathcal{C}\Big\}$$

the **outer** or **exterior Lebesgue measure** of Γ. What we are going to do is describe a family $\overline{\mathcal{B}}_{\mathbf{R}^N}$ for which the map

$$\Gamma \in \overline{\mathcal{B}}_{\mathbf{R}^N} \longmapsto |\Gamma|_e$$

satisfies (II.0.4). (The notation here, in particular the *bar*, will be explained in ii) of Examples III.1.5 and Exercise III.1.9.) However, before starting on this project, we first check that $|I|_e = \text{vol}(I)$ for rectangles I.

II.1.1 Lemma. *If $\Gamma = \bigcup_1^n J_m$ where the J_m's are non-overlapping rectangles, then $|\Gamma|_e = \sum_1^n \text{vol}(J_m)$.*

PROOF: Obviously $|\Gamma|_e \leq \sum_1^n \text{vol}(J_m)$. To prove the opposite inequality, let $\mathcal{C}$ be a cover of J. Given an $\epsilon > 0$, choose $\tilde{I}$ for each $I \in \mathcal{C}$ so that $I \subseteq \tilde{I}^\circ$ and $\text{vol}(\tilde{I}) \leq (1 + \epsilon)\text{vol}(I)$. Because Γ is compact, there exist $\{I_1, \ldots, I_L\} \subseteq \mathcal{C}$ such that $\Gamma \subseteq \bigcup_1^L \tilde{I}_\ell$. In particular, by Lemma II.1.1,

$$\sum_{m=1}^n \text{vol}(J_m) \leq \sum_{m=1}^n \sum_{\ell=1}^L \text{vol}(J_m \cap \tilde{I}_\ell)$$

$$\leq \sum_{\ell=1}^L \text{vol}(\tilde{I}_\ell) \leq (1 + \epsilon) \sum_\ell^L \text{vol}(I_\ell) \leq (1 + \epsilon)\Sigma(\mathcal{C}).$$

(In the preceding, we have used the fact that, for any pair of rectangles I and J, either $I \cap J = \emptyset$, in which case we take $\text{vol}(I \cap J) \equiv 0$, or $I \cap J$ is itself a rectangle.) After letting $\epsilon \searrow 0$ and then taking the infimum over $\mathcal{C}$'s, we get the required result. ∎

In view of Lemma II.1.1, we are justified in replacing $\text{vol}(I)$ by $|I|_e$ for rectangles I.

Our next result shows that half the equality in (II.0.4) is automatic, even before we restrict to Γ's from $\overline{\mathcal{B}}_{\mathbf{R}^N}$.

II.1.2 Lemma. *If $\Gamma_1 \subseteq \Gamma_2$, then $|\Gamma_1|_e \leq |\Gamma_2|_e$. In addition, if $\Gamma \subseteq \bigcup_1^\infty \Gamma_n$, then $|\Gamma|_e \leq \sum_1^\infty |\Gamma_n|_e$. In particular, if $\Gamma \subseteq \bigcup_1^\infty \Gamma_n$ and $|\Gamma_n|_e = 0$ for each $n \geq 1$, then $|\Gamma|_e = 0$; and so $|\partial I|_e = 0$ for any rectangle I. (Here, and throughout, ∂S will denote the boundary $\overline{S} \setminus S^\circ$ of the set S.)*

PROOF: The first assertion follows immediately from the fact that every cover of Γ_1 is also a cover of Γ_2.

In order to prove the second assertion, let $\epsilon > 0$ be given, and choose for each $n \geq 1$ a cover $\mathcal{C}_n$ so that $\sum_{I \in \mathcal{C}_n} |I|_e \leq |\Gamma_n|_e + \epsilon/2^n$. It is obvious that $\mathcal{C} \equiv \bigcup_1^\infty \mathcal{C}_n$ is a countable cover of Γ. Hence

$$|\Gamma|_e \leq \sum_{I \in \mathcal{C}} |I|_e \leq \sum_{n=1}^\infty \sum_{I \in \mathcal{C}_n} \mathrm{vol}\,(I) \leq \sum_{n=1}^\infty (|\Gamma_n|_e + \epsilon/2^n) \leq \sum_{n=1}^\infty |\Gamma_n|_e + \epsilon.$$

Given the preceding, it is clear that the only point which still requires comment is the proof that, for any $1 \leq k \leq N$ and Γ of the form $[a_1, b_1] \times \cdots \times \{c_k\} \times \cdots \times [a_N, b_N]$, $|\Gamma|_e = 0$. But obviously such a $\Gamma \subseteq I_\delta$ for every $\delta > 0$, where $I_\delta = [a_1, b_1] \times \cdots \times [c_k - \delta, c_k + \delta] \times \cdots \times [a_N, b_N]$; and so $|\Gamma|_e \leq |I_\delta|_e \leq K\delta$ for some $K < \infty$ and all $\delta > 0$. ∎

II.1.3 Remark.

We will use $\mathfrak{G}$ to denote the class of all open subsets in a topological space. Thus, in the present context, $\mathfrak{G}$ stands for the class of open subsets of $\mathbb{R}^N$. In this connection, we also introduce the class $\mathfrak{G}_\delta$ which consists of all subsets which can be written as the intersection of a countable number of open subsets. Note that $\mathfrak{G} \cup \mathfrak{F} \subseteq \mathfrak{G}_\delta$, where we use $\mathfrak{F}$ to stand for the class of all closed subsets. Finally, note that $\Gamma \in \mathfrak{G}_\delta$ if and only if its complement $\Gamma^{\complement}$ is an element of $\mathfrak{F}_\sigma$, the class of subsets which can be written as the union of a countable number of closed sets.

II.1.4 Lemma. *For any $\Gamma \subseteq \mathbb{R}^N$*

(II.1.5) $$|\Gamma|_e = \inf\big\{|G|_e : \Gamma \subseteq G \in \mathfrak{G}\big\}.$$

In particular, for each $\Gamma \subseteq \mathbb{R}^N$ there is an $B \in \mathfrak{G}_\delta$ such that $\Gamma \subseteq B$ and $|\Gamma|_e = |B|_e$.

PROOF: Obviously the left hand side of (II.1.5) is dominated by the right hand side. To prove the opposite inequality, let $\epsilon > 0$ be given, and choose $\mathcal{C} = \{I_n\}_1^\infty$ to be a cover of Γ for which $|\Gamma|_e \geq \Sigma(\mathcal{C}) - \epsilon/2$. Next, for each $n \geq 1$, let $\tilde{I}_n$ be a rectangle satisfying $I \subseteq \tilde{I}_n^\circ$ and $|\tilde{I}_n|_e \leq |I_n|_e + \epsilon/2^{n+1}$. Then $G \equiv \bigcup_1^\infty \tilde{I}_n^\circ$ is certainly open, it contains Γ, and

$$|G|_e \leq \sum_1^\infty |\tilde{I}_n|_e \leq |\Gamma|_e + \epsilon.$$

Having proved the first assertion, the second one follows by choosing a sequence $\{G_n\}_1^\infty \subseteq \mathfrak{G}$ so that $\Gamma \subseteq G_n$ and $|G_n|_e \leq |\Gamma|_e + \frac{1}{n}$ for each $n \geq 1$. Clearly the set $B \equiv \bigcap_1^\infty G_n$ will then serve. ∎

We are now ready to describe the class $\overline{\mathcal{B}}_{\mathbb{R}^N}$ (alluded to at the beginning of this section), although it will not be immediately clear why it has the properties which we want. Be that as it may, we will say that $\Gamma \subseteq \mathbb{R}^N$ is **Lebesgue measurable** (or, when it is clear that we are discussing LEBESGUE's measure, simply **measurable**) and we will write $\Gamma \in \overline{\mathcal{B}}_{\mathbb{R}^N}$ if for each $\epsilon > 0$ there is an open $G \supseteq \Gamma$ such that $|G \setminus \Gamma|_e < \epsilon$. In order to distinguish $|\cdot|_e$ from its restriction to $\overline{\mathcal{B}}_{\mathbb{R}^N}$, we will use $|\Gamma|$ instead of $|\Gamma|_e$ when Γ is measurable, and we will call $|\Gamma|$ the **Lebesgue's measure** (or simply, the **measure**) of Γ.

II.1.6 Remark.

At first sight one might be tempted to say that, in view of Lemma II.1.4, every subset Γ is measurable. This is because one is inclined to think that $|G|_e = |G \setminus \Gamma|_e + |\Gamma|_e$ when, in fact, $|G|_e \leq |G \setminus \Gamma|_e + |\Gamma|_e$ is all that we know. Therein lies the subtlety of the definition! Nonetheless, it is clear that every open G is measurable. Furthermore, if $|\Gamma|_e = 0$, then Γ is measurable since we can choose, for any $\epsilon > 0$, an open $G \supseteq \Gamma$ such that $|G \setminus \Gamma|_e \leq |G| < \epsilon$. Finally, if Γ is measurable, then there is a $B \in \mathcal{G}_\delta$ such that $\Gamma \subseteq B$ and $|B \setminus \Gamma|_e = 0$. Indeed, simply choose $\{G_n\}_1^\infty \subseteq \mathcal{G}$ so that $\Gamma \subseteq G_n$ and $|G_n \setminus \Gamma|_e < 1/n$, and take $B = \bigcap_1^\infty G_n$.

Our next result shows that lots of sets are measurable.

II.1.7 Lemma. *If $\{\Gamma_n\}_1^\infty$ is a sequence of measurable sets, then $\Gamma = \bigcup_1^\infty \Gamma_n$ is also measurable and, of course,*

$$(\text{II.1.8}) \qquad |\Gamma| \leq \sum_1^\infty |\Gamma_n|.$$

PROOF: For each $n \geq 1$, choose $\Gamma \subseteq G_n \in \mathcal{G}$ so that $|G_n \setminus \Gamma_n|_e < \epsilon/2^n$. Then $G \equiv \bigcup_1^\infty G_n$ is open, contains Γ, and satisfies

$$|G \setminus \Gamma|_e \leq \left| \bigcup_1^\infty (G_n \setminus \Gamma_n) \right|_e \leq \sum_1^\infty |G_n \setminus \Gamma_n|_e < \epsilon. \quad \blacksquare$$

Writing a rectangle $I = I^\circ \cup \partial I$, we now see that every rectangle is measurable.

Knowing that $\overline{\mathcal{B}}_{\mathbb{R}^N}$ is closed under countable unions, our next goal is to prove that it is also closed under complementation. In doing so, we will be simultaneously coming closer to showing that (II.0.4) holds for $|\cdot|_e$ on $\overline{\mathcal{B}}_{\mathbb{R}^N}$. We begin with a general fact about $|\cdot|_e$.

II.1.9 Lemma. *If the distance*

$$\text{dist}\,(\Gamma_1, \Gamma_2) \equiv \inf\{|y - x| : x \in \Gamma_1 \text{ and } y \in \Gamma_2\}$$

is positive, then $|\Gamma_1 \cup \Gamma_2|_e = |\Gamma_1|_e + |\Gamma_2|_e$.

PROOF: We need only show that $|\Gamma_1 \cup \Gamma_2|_e \geq |\Gamma_1|_e + |\Gamma_2|_e$. To this end, let $\epsilon > 0$ be given, and choose $\mathcal{C}$ to be a cover of $\Gamma_1 \cup \Gamma_2$ with

$$\left|\Gamma_1 \cup \Gamma_2\right|_e + \epsilon \geq \Sigma(\mathcal{C}).$$

Without loss in generality, we assume that $\delta \equiv \text{diam}(I) < \text{dist}(\Gamma_1, \Gamma_2)$ for each $I \in \mathcal{C}$. (Indeed, if this is not already true, one can repeatedly subdivide any I until one gets rectangles whose diameters are shorter than δ.) Hence, if

$$\mathcal{C}_i \equiv \left\{ I \in \mathcal{C} : I \cap \Gamma_i \neq \emptyset \right\}, \quad i \in \{1, 2\},$$

then $\mathcal{C}_1 \cap \mathcal{C}_2 = \emptyset$, $\mathcal{C} \supseteq \mathcal{C}_1 \cup \mathcal{C}_2$, and $\mathcal{C}_i$ covers Γ_i. In particular,

$$\left|\Gamma_1 \cup \Gamma_2\right|_e + \epsilon \geq \Sigma(\mathcal{C}) \geq \Sigma(\mathcal{C}_1) + \Sigma(\mathcal{C}_2) \geq |\Gamma_1|_e + |\Gamma_2|_e. \ \blacksquare$$

A **cube** Q in $\mathbb{R}^N$ is a rectangle all of whose sides have the same length. The next easy covering lemma is the final ingredient which we need before proving that $\overline{\mathcal{B}}_{\mathbb{R}^N}$ is closed under complementation.

II.1.10 Lemma. *If G is an open set in $\mathbb{R}$, then G is the union of a countable number of mutually disjoint open intervals. More generally, if G is an open set in $\mathbb{R}^N$, then, for each $\delta > 0$, G admits a countable, non-overlapping, exact cover $\mathcal{C}$ by cubes Q with* $\text{diam}\,(Q) < \delta$.

PROOF: If $G \subseteq \mathbb{R}$ is open and $x \in G$, let I_x be the open connected component of G containing x. Then I_x is an open interval and, for any $x, y \in G$, either $I_x \cap I_y = \emptyset$ or $I_x = I_y$. Hence, $\mathcal{C} \equiv \{I_x : x \in G \cap \mathbb{Q}\}$ ($\mathbb{Q}$ denotes the set of rational numbers) is the required cover.

To handle the general case, set $Q_n = [0, 1/2^n]^N$ and $\mathcal{K}_n = \{\frac{\mathbf{k}}{2^n} + Q_n : \mathbf{k} \in \mathbb{Z}^N\}$. Note that if $m \leq n$, $Q \in \mathcal{K}_m$, and $\hat{Q} \in \mathcal{K}_n$, then either $\hat{Q} \subseteq Q$ or $Q^\circ \cap \hat{Q}^\circ = \emptyset$. Now let $G \subseteq \mathbb{R}^N$ and $\delta > 0$ be given. Let n_0 be the smallest $n \in \mathbb{Z}$ such that $\frac{N^{1/2}}{2^n} < \delta$, and set $\mathcal{C}_{n_0} = \{Q \in \mathcal{K}_{n_0} : Q \subseteq G\}$. Next, define $\mathcal{C}_n$ inductively for $n > n_0$ so that $\mathcal{C}_{n+1} = \{Q \in \mathcal{K}_{n+1} : Q \subseteq H_n\}$, where $H_n = G\backslash \left(\bigcup_{m=n_0}^n (\bigcup \mathcal{C}_m)^\circ \right)$. Note that if $m \leq n$, $Q \in \mathcal{C}_m$, and $\hat{Q} \in \mathcal{C}_n$, then either $Q = \hat{Q}$ or $Q^\circ \cap \hat{Q}^\circ = \emptyset$. Hence $\mathcal{C} \equiv \bigcup_{n=n_0}^\infty \mathcal{C}_n$ is non-overlapping, and certainly $\bigcup \mathcal{C} \subseteq G$. Finally, if $x \in G$, choose $n \geq n_0$ and $\hat{Q} \in \mathcal{K}_n$ so that $x \in \hat{Q} \subseteq G$. If $\hat{Q} \notin \mathcal{C}_n$, then there is an $n_0 \leq m < n$ and a $Q \in \mathcal{C}_m$ such that $Q^\circ \cap \hat{Q}^\circ \neq \emptyset$. But this means that $\hat{Q} \subseteq Q$ and therefore that $x \in Q \subseteq \bigcup \mathcal{C}$. Thus $\mathcal{C}$ covers G. $\blacksquare$

II.1.11 Lemma. *If Γ is measurable then so is its complement $\Gamma^{\complement}$.*

PROOF: We first check that every compact set K is measurable. To this end, let $\epsilon > 0$ be given and choose an open set $G \supseteq K$ so that $|G| - |K|_e < \epsilon$. Set $H = G \setminus K$ and choose a non-overlapping sequence $\{Q_n\}_1^{\infty}$ of cubes for which $H = \bigcup_1^{\infty} Q_n$. By Lemma II.1.1, $\sum_1^n |Q_m| = |\bigcup_1^n Q_m|$. Moreover, since K and $\bigcup_1^n Q_m$ are disjoint compact sets, Lemma II.1.9 says that $\left|\left(\bigcup_1^n Q_m\right) \cup K\right|_e = |\bigcup_1^n Q_m| + |K|_e$. Hence

$$|G| \geq \left|\left(\bigcup_1^n Q_m\right) \cup K\right|_e = \left|\bigcup_1^n Q_m\right| + |K|_e = \sum_1^n |Q_m| + |K|_e,$$

and so $\sum_1^n |Q_m| \leq |G| - |K|_e < \epsilon$ for all $n \geq 1$. As a consequence, we now see that $|H|_e \leq \sum_1^{\infty} |Q_m| \leq \epsilon$; and so K is measurable.

We next show that every closed set F is measurable. To this end, simply write $F = \bigcup_1^{\infty} \left(F \cap \overline{B(0,n)}\right)$, where $B(x,r)$ denotes the ball $\{y \in \mathbb{R}^N : |y - x| < r\}$ of radius r around the point x. Since each $F \cap \overline{B(0,n)}$ is compact, it follows from the preceding and Lemma II.1.7 that F is measurable.

To complete the proof, first observe that we now know that $\mathcal{F}_{\sigma} \subseteq \overline{\mathcal{B}}_{\mathbb{R}^N}$. Next (cf. Remark II.1.6) choose $B \in \mathfrak{G}_{\delta}$ so that $\Gamma \subseteq B$ and $|B \setminus \Gamma|_e = 0$. Then, since $B^{\complement} \in \mathfrak{F}_{\sigma}$ and $|B \setminus \Gamma|_e = 0$, $\Gamma^{\complement} = B^{\complement} \cup (B \setminus \Gamma)$ is measurable. ∎

We have at last arrived at our goal.

II.1.12 Theorem. *The class $\overline{\mathcal{B}}_{\mathbb{R}^N}$ contains $\mathfrak{G}$, is closed under countable unions, complementation, and therefore also under differences and countable intersections. Hence, $\mathfrak{G}_{\delta} \cup \mathfrak{F}_{\sigma} \subseteq \overline{\mathcal{B}}_{\mathbb{R}^N}$; in fact, $\Gamma \in \overline{\mathcal{B}}_{\mathbb{R}^N}$ if and only if there exist $A \in \mathfrak{F}_{\sigma}$ and $B \in \mathfrak{G}_{\delta}$ such that $A \subseteq \Gamma \subseteq B$ and $|B \setminus A| = 0$, in which case $|\Gamma| = |A| = |B|$. Finally, if $\{\Gamma_n\}_1^{\infty} \subseteq \overline{\mathcal{B}}_{\mathbb{R}^N}$ is a sequence of sets, then*

$$(\text{II.1.13}) \qquad \left|\bigcup_1^{\infty} \Gamma_n\right| = \sum_1^{\infty} |\Gamma_n| \qquad if \quad \Gamma_m \cap \Gamma_n = \emptyset \quad for \quad m \neq n.$$

PROOF: The first assertion follows immediately from what we already know together with trivial manipulations of set theoretic operations; and clearly the fact that $\mathfrak{G}_{\delta} \cup \mathcal{F}_{\sigma} \subseteq \overline{\mathcal{B}}_{\mathbb{R}^N}$ is a consequence of the first assertion. Next suppose that Γ is measurable. By the final part of Remark II.1.6 applied to Γ and $\Gamma^{\complement}$, we can find $A \in \mathfrak{F}_{\sigma}$ and $B \in \mathfrak{G}_{\delta}$ such that $\Gamma^{\complement} \subseteq A^{\complement}$, $\Gamma \subseteq B$, and $|B \setminus \Gamma| = |A^{\complement} \setminus \Gamma^{\complement}| = 0$. On the other hand, if there exist $A \in \mathcal{F}_{\sigma}$ and $B \in \mathfrak{G}_{\delta}$ such that $A \subseteq \Gamma \subseteq B$ and $|B \setminus A| = 0$, then $\Gamma = A \cup (\Gamma \setminus A)$ is measurable because $|\Gamma \setminus A|_e \leq |B \setminus A| = 0$. Hence, it remains only to check (II.1.13).

We first prove (II.1.13) under the additional assumption that each of the Γ_n 's is bounded. Given $\epsilon > 0$, choose open sets G_n so that $\Gamma_n^0 \subseteq G_n$ and $|G_n \setminus \Gamma_n^0| < \epsilon/2^n$. Then $K_n \equiv G_n^0 \subseteq \Gamma_n$ is compact and $|\Gamma_n \setminus K_n| < \epsilon/2^n$. Since $K_m \cap K_n = \emptyset$ and therefore $\mathrm{dist}(K_m, K_n) > 0$ for $m \neq n$, we have, by Lemma II.1.9, that $|\bigcup_1^n K_m| = \sum_1^n |K_m|$ for every $n \geq 1$. Hence

$$\sum_{m=1}^\infty |\Gamma_m| < \sum_{m=1}^\infty |K_m| + \epsilon = \lim_{n \to \infty} \left| \bigcup_{m=1}^n K_m \right| + \epsilon \leq \left| \bigcup_{m=1}^\infty \Gamma_m \right| + \epsilon.$$

That is, $\sum_1^\infty |\Gamma_m| \leq |\bigcup_1^\infty \Gamma_m|$. Since the opposite inequality always holds, (II.1.13) is now proved for bounded Γ_n 's.

Finally, to handle the general case, set $A_1 = B(0,1)$ and $A_{n+1} = B(0, n+1) \setminus B(0,n)$. Then

$$(\Gamma_m \cap A_n) \cap (\Gamma_{m'} \cap A_{n'}) = \emptyset \quad \text{if} \quad (m,n) \neq (m', n').$$

Hence, by the preceding,

$$\sum_{m=1}^\infty |\Gamma_m| = \sum_{m=1}^\infty \sum_{n=1}^\infty |\Gamma_m \cap A_n| = \sum_{n=1}^\infty \sum_{m=1}^\infty |\Gamma_m \cap A_n|$$

$$= \sum_{n=1}^\infty \left| \bigcup_{m=1}^\infty (\Gamma_m \cap A_n) \right| = \sum_{n=1}^\infty \left| \left(\bigcup_{m=1}^\infty \Gamma_m \right) \cap A_n \right|$$

$$= \left| \bigcup_{n=1}^\infty \left[\left(\bigcup_{m=1}^\infty \Gamma_m \right) \cap A_n \right] \right| = \left| \bigcup_{m=1}^\infty \Gamma_m \right|. \quad \blacksquare$$

II.1.14 Remark.

Although it seems hardly necessary to point out, outer Lebesgue measure has an obvious but extremely important property: it is **invariant under translation.** That is, $|x + \Gamma|_e = |\Gamma|_e$ for all $x \in \mathbb{R}^N$ and all $\Gamma \subseteq \mathbb{R}^N$. As a consequence, we also see that $x + \Gamma$ is measurable whenever $x \in \mathbb{R}^N$ and Γ itself is measurable.

Before concluding this preliminary discussion of Lebesgue's measure, it may be appropriate to examine whether there are any non-measurable sets. It turns out that the existence of non-measurable sets brings up some extremely delicate points about the foundations of mathematics. Indeed, if one is willing to abandon the full axiom of choice, then R. Solovay has shown that there is a model of mathematics in which *every* subset of $\mathbb{R}^N$ is LEBESGUE measurable. However, if one accepts the full axiom of choice, then the following argument, due to VITALI, shows that there are sets which are not LEBESGUE measurable. The use of the axiom of choice comes in Lemma II.1.16 below, it is not used in the proof of the next lemma, a result which is interesting in its own right.

II.1.15 Lemma. *If* Γ *is a measurable subset of* $\mathbb{R}$ *and* $|\Gamma| > 0$, *then the set* $\Gamma - \Gamma \equiv \{y - x : x, y \in \Gamma\}$ *contains the open interval* $(-\delta, \delta)$ *for some* $\delta > 0$.

PROOF: Without loss in generality, we assume that $|\Gamma| < \infty$.

Choose an open set $G \supseteq \Gamma$ so that $|G \setminus \Gamma| < \frac{1}{3}|\Gamma|$, and let $\mathcal{C}$ be a countable collection of mutually disjoint, non-empty, open intervals I whose union is G (cf. Lemma II.1.10). Then

$$\sum_{I \in \mathcal{C}} |I \cap \Gamma| = |\Gamma| \geq \frac{3}{4}|G| = \frac{3}{4}\sum_{I \in \mathcal{C}} |I|.$$

Hence, there must be an $I \in \mathcal{C}$ such that $|I \cap \Gamma| \geq \frac{3}{4}|I|$. Set $A = I \cap \Gamma$. If $d \in \mathbb{R}$ and $(d + A) \cap A = \emptyset$, then

$$2|A| = |d + A| + |A| = |(d + A) \cup A| \leq |(d + I) \cup I|.$$

At the same time, $(d+I) \cup I \subseteq (I^-, d+I^+)$ if $d \geq 0$ and $(d+I) \cup I \subseteq (d+I^-, I^+)$ if $d < 0$; and so, in either case, $|(d+I) \cup I| \leq |d| + |I|$. Hence, if $(d+A) \cap A = \emptyset$, then $2|A| \leq |d| + |I|$, from which we deduce that $|d| \geq \frac{1}{2}|I|$. In other words, if $|d| < \frac{1}{2}|I|$, then $(d+A) \cap A \neq \emptyset$. But this means that for every $d \in \left(-\frac{1}{2}|I|, \frac{1}{2}|I|\right)$ there exist $x, y \in A \subseteq \Gamma$ such that $d = y - x$. ∎

II.1.16 Lemma. *Assuming the axiom of choice, there is a subset* A *of* $\mathbb{R}$ *such that* $(A - A) \cap \mathbb{Q} = \{0\}$ *and yet* $\mathbb{R} = \bigcup_{q \in \mathbb{Q}}(q + A)$.

PROOF: Write $x \sim y$ if $y - x \in \mathbb{Q}$. Then "$\sim$" is an equivalence relation on $\mathbb{R}$, and for each $x \in \mathbb{R}$ the equivalence class $[x]^\sim$ of x is $x + \mathbb{Q}$. Now, using the axiom of choice, let A be a set which contains precisely one element from each of the equivalence classes $[x]^\sim$, $x \in \mathbb{R}$. It is then clear that A has the required properties. ∎

II.1.17 Theorem. *Assuming the axiom of choice, every* $\Gamma \subseteq \mathbb{R}$ *with* $|\Gamma|_e > 0$ *contains a non-measurable subset.*

PROOF: Let A be the set constructed in Lemma II.1.16. Then $0 < |\Gamma|_e = \sum_{q \in \mathbb{Q}} |\Gamma \cap (q + A)|_e$, and so there must exist a $q \in \mathbb{Q}$ such that $|\Gamma \cap (q+A)|_e > 0$.

Hence, if $\Gamma \cap (q + A)$ were measurable, then, by Lemma II.1.15, we would have that $(-\delta, \delta) \subseteq \{y - x : x, y \in (q + A)\} \subseteq \{0\} \cup \mathbb{Q}^\delta$ for some $\delta > 0$. ∎

II.1.18 Exercise.

Let Γ_1 and Γ_2 be measurable subsets in $\mathbb{R}^N$. If $\Gamma_1 \subseteq \Gamma_2$ and $|\Gamma_1| < \infty$, show that $|\Gamma_2 \setminus \Gamma_1| = |\Gamma_2| - |\Gamma_1|$. More generally, show that if $|\Gamma_1 \cap \Gamma_2| < \infty$, then $|\Gamma_1 \cup \Gamma_2| = |\Gamma_1| + |\Gamma_2| - |\Gamma_1 \cap \Gamma_2|$.

II.1.19 Exercise.

Let $\{\Gamma_n\}_1^\infty$ be a sequence of measurable sets in $\mathbb{R}^N$. Assuming that $|\Gamma_m \cap \Gamma_n| = 0$ for $m \neq n$, show that $|\bigcup_1^\infty \Gamma_n| = \sum_1^\infty |\Gamma_n|$.

II.1.20 Exercise.

It is clear that any countable set has LEBESGUE measure zero. However, it is not so immediately clear that there are uncountable subsets of $\mathbb{R}$ whose LEBESGUE measure is zero. We will show here how to construct such a set. Namely, start with the set $C_0 = [0, 1]$ and let C_1 be the set obtained by removing the open middle third of C_0 (i.e., $C_1 = C_0 \setminus (1/3, 2/3) = [0, 1/3] \cup [2/3, 1]$). Next, let C_2 be the set obtained from C_1 after removing the open middle third of each of the (two) intervals of which C_1 is the disjoint union. More generally, given C_k (which is the union of 2^k disjoint, closed intervals), let C_{k+1} be the set which one gets from C_k by removing the open middle third of each of the intervals of which C_k is the disjoint union. Finally, set $C = \bigcap_{k=0}^\infty C_k$. The set C is called the **Cantor set**, and it turns out to be an extremely useful source of examples. Here we will show that it is an example of an uncountable set of LEBESGUE measure zero.

i) Note that C is closed and that $|C| \leq |C_k| = (2/3)^k$, $k \geq 0$. Conclude that $C \in \overline{\mathcal{B}}_\mathbb{R}$ and that $|C| = 0$.

ii) Let $\mathcal{A}$ denote the set of $\alpha \in \{0, 1, 2\}^\mathbb{N}$ with the properties that:

   a) $\alpha_0 \in \{0, 1\}$ and $\alpha_0 = 1$ only if $\alpha_k = 0$ for all $k \in \mathbb{Z}^+$;

   b) $\alpha_k \in \{0, 1\}$ for infinitely many $k \in \mathbb{Z}^+$.

Check that the map

$$\alpha \in \mathcal{A} \longmapsto \sum_{k \in \mathbb{N}} \frac{\alpha_k}{3^k} \in [0, 1]$$

is an one-to-one and onto; and let $x \in [0, 1] \longmapsto \alpha(x) \in \mathcal{A}$ denote the inverse mapping. Next, define $\mathcal{A}_0$ to be the set of $\alpha \in \mathcal{A}$ such that $\alpha_k = 0$ for all but a finite number of $k \in \mathbb{N}$; and, for $\alpha \in \mathcal{A}_0$, define $\ell(\alpha) = \max\{k \in \mathbb{N} : \alpha_k \neq 0\}$. Show that

$$\partial C_\ell = \Big\{ x : \alpha(x) \in \mathcal{A}_0, \; \ell(\alpha(x)) = \ell, \text{ and } \alpha_k(x) \in \{0,2\} \text{ for } 0 \leq k < \ell \Big\};$$

and conclude that

$$C_\ell^\circ = \Big\{ x : \alpha_k(x) \in \{0,2\} \text{ for every } 0 \leq k \leq \ell$$
$$\text{and } \alpha_k(x) \neq 0 \text{ for some } k > \ell \Big\}.$$

Finally, define

$$\mathcal{A}^\circ = \Big\{ \alpha \in \mathcal{A} \setminus \mathcal{A}_0 : \alpha_k \in \{0,2\} \text{ for all } k \in \mathbb{N} \Big\},$$

and show that

$$\bigcap_{\ell=0}^{\infty} C_\ell^\circ = \Big\{ x : \alpha(x) \in \mathcal{A}^\circ \Big\}$$

while

$$C \setminus \bigcap_{\ell=0}^{\infty} C_\ell^\circ = \Big\{ x : \alpha(x) \in \mathcal{A}_0 \text{ and } \alpha_k(x) \in \{0,2\} \text{ for every } 0 \leq k < \ell(\alpha) \Big\}.$$

iii) To see that C is not countable, suppose that it were. Using ii), show that one would then have that $\{0,2\}^{\mathbb{Z}^+}$ is countable. Finally, recall CAN-TOR's famous *anti-diagonalization procedure* for showing that $\{0,2\}^{\mathbb{Z}^+}$ cannot be counted.

II.2. Euclidean Invariance.

Although the property of translation invariance was built into our construction of LEBESGUE's measure, it is not immediately obvious how LEBESGUE's measure reacts to rotations of $\mathbb{R}^N$. One suspects that, as *the natural* measure on $\mathbb{R}^N$, LEBESGUE's measure should be invariant under the full group of EUCLIDean transformations (i.e., rotations as well as translations). However, because our definition of the measure was based on rectangles and the rectangles were inextricably tied to a fixed set of coordinate axes, rotation invariance is not so clear as translation invariance. In the present section we will see how LEBESGUE's measure transforms under an arbitrary linear transformation of $\mathbb{R}^N$, and rotation invariance will follow as an immediate corollary.

We begin with a results about the behavior of measurable sets under general transformations.

II.2.1 Lemma. *Let $F \subseteq \mathbb{R}^N$ be closed and $\Phi : F \longrightarrow \mathbb{R}^N$ continuous. Then $\Phi(\Gamma \cap F) \in \mathfrak{F}_\sigma$ whenever $\Gamma \in \mathfrak{F}_\sigma$. Furthermore, if in addition, $|\Phi(\Gamma \cap F)|_e = 0$ whenever $|\Gamma|_e = 0$, then $\Phi(\Gamma \cap F)$ is measurable whenever Γ is. In particular, if Φ is* LIPSCHITZ *continuous with* LIPSCHITZ *constant L (i.e., $|\Phi(y) - \Phi(x)| \leq L|y - x|$ for all $x, y \in F$) then $|\Phi(\Gamma \cap F)|_e \leq L^N |\Gamma|_e$ and therefore Φ takes measurable subsets of F into measurable sets.*

PROOF: Remember that functions preserve unions. Hence, the class of sets Γ for which $\Phi(\Gamma \cap F) \in \mathfrak{F}_\sigma$ is closed under countable unions. Next note that if K is compact, then, by continuity, so is $\Phi(K \cap F)$. But every closed set in $\mathbb{R}^N$ is the countable union of compact sets, and therefore we see that $\Phi(\Gamma \cap F) \in \mathfrak{F}_\sigma$ for every closed Γ. Finally, since every $\Gamma \in \mathfrak{F}_\sigma$ is a countable union of closed sets, the first assertion is proved.

Next assume, in addition, that $|\Phi(\Gamma \cap F)|_e = 0$ whenever $|\Gamma|_e = 0$. Given a measurable Γ, choose $A \in \mathcal{F}_\sigma$ so that $A \subseteq \Gamma$ and $|\Gamma \setminus A| = 0$. Then $\Phi(\Gamma \cap F) = \Phi(A \cap F) \cup \Phi((\Gamma \setminus A) \cap F)$ is measurable because $\Phi(A \cap F) \in \mathfrak{F}_\sigma$ and $\big|\Phi((\Gamma \setminus A) \cap F)\big|_e = 0$.

We now show that if Φ is LIPSCHITZ continuous with LIPSCHITZ constant L, then $|\Phi(\Gamma \cap F)|_e \leq L^N |\Gamma|_e$. But clearly it suffices to do this when Γ is a rectangle $I = \prod_1^N [c_k - \delta_k, c_k + \delta_k]$, in which case $\Phi(I \cap F) \subseteq \prod_1^N [\Phi(c_k) - L\delta_k, \Phi(c_k) + L\delta_k]$; and so $|\Phi(I \cap F)| \leq L^N |I|$. ∎

Given an $N \times N$ matrix A of real numbers a_{ij}, we will use T_A to denote the linear transformation of $\mathbb{R}^N$ which A determines relative to the standard basis $\{e_1, \ldots, e_N\}$. That is,

$$T_A x = \begin{bmatrix} \sum_{j=1}^N a_{1j} x^j \\ \vdots \\ \sum_{j=1}^N a_{Nj} x^j \end{bmatrix}, \qquad \text{for } x = \begin{bmatrix} x^1 \\ \vdots \\ x^N \end{bmatrix} \in \mathbb{R}^N$$

Since T_A is obviously LIPSCHITZ continuous, T_A takes measurable sets into measurable sets. The main result of this section is the following important fact about LEBESGUE's measure.

II.2.2 Theorem. *Given a real $N \times N$ matrix A, T_A takes measurable sets into measurable sets and $|T_A(\Gamma)|_e = |\det(A)||\Gamma|_e$ for all $\Gamma \subseteq \mathbb{R}^N$. (We use $\det(A)$ to denote the determinant of A.)*

PROOF: There are several steps.

Step 1). *For any $c \in \mathbb{R}$, $\lambda \geq 0$, and $\Gamma \subseteq \mathbb{R}^N$, $|(c + \lambda\Gamma)|_e = |\lambda|^N |\Gamma|_e$, where $\lambda\Gamma \equiv \{\lambda x : x \in \Gamma\}$.*

By translation invariance, we may and will assume that $c = 0$. Moreover, there is nothing to prove when $\lambda = 0$. Finally, it is clear that $|\lambda I| = |\lambda|^N |I|$ for any $\lambda \neq 0$ and rectangle I. Hence, since $\mathcal{C}$ is a countable cover of Γ by rectangles I if and only if $\{\lambda I : I \in \mathcal{C}\}$ is a countable cover of $\lambda\Gamma$, we are done.

Step 2). *For any linear transformation T and all cubes Q, $|T(Q)| = \alpha(T)|Q|$ where $\alpha(T) \equiv |T(Q_0)|$ and $Q_0 = [0, 1]^N$.*

Since every cube $Q = c + \lambda Q_0$ for some $c \in \mathbb{R}^N$ and λ satisfying $|\lambda|^N = |Q|$, Step 1) plus the linearity of T yields $|T(Q)| = |(T(c) + \lambda T(Q_0)| = |\lambda|^N |T(Q_0)| = \alpha(T)|Q|$.

Step 3). *For any linear transformation T and open G,*

$$|T(G)| \leq \alpha(T)|G|.$$

Moreover, equality holds if T is non-singular.

Let $\mathcal{C}$ be a countable, exact cover of G by non-overlapping cubes Q. Then

$$|T(G)| \leq \sum_{Q \in \mathcal{C}} |T(Q)| = \alpha(T) \sum_{Q \in \mathcal{C}} |Q| = \alpha(T)|G|.$$

Now suppose that T is non-singular. Then

$$T(Q) \setminus T(Q^\circ) = T(Q \setminus Q^\circ)$$

has measure 0, and

$$T(Q^\circ) \cap T((Q')^\circ) = \emptyset \quad \text{for distinct} \quad Q, Q' \in \mathcal{C}.$$

Hence $|T(Q)| = |T(Q^\circ)|$ and

$$|T(G)| \geq \sum_{Q \in \mathcal{C}} |T(Q^\circ)| = \sum_{Q \in \mathcal{C}} |T(Q)| = \alpha(T) \sum_{Q \in \mathcal{C}} |Q| = \alpha(T)|G|.$$

Step 4). *For any non-singular linear transformation T and all $\Gamma \subseteq \mathbb{R}^N$, $|T(\Gamma)|_e = \alpha(T)|\Gamma|_e$.*

Since $\Gamma \subseteq G \in \mathfrak{G}$ if and only if $T(\Gamma) \subseteq T(G) \in \mathfrak{G}$, this step is an immediate consequence of (II.1.5) and Step 3).

Step 5). *If S and T are non-singular linear transformations, then $\alpha(S \circ T) = \alpha(S)\alpha(T)$.*

Simply note that, by Step 4):

$$\alpha(S \circ T) = |S \circ T(Q_0)| = |S(T(Q_0))| = \alpha(S)|T(Q_0)| = \alpha(S)\alpha(T).$$

Step 6). *If A is an orthogonal matrix, then $\alpha(T_A) = 1$.*

Because A is orthogonal, $B(0,1) = T_A(B(0,1))$ and therefore $|B(0,1)| = \alpha(T_A)|B(0,1)|$.

Step 7). *If A is non-singular and symmetric, then $\alpha(T_A) = |\det(A)|$.*

If A is already diagonal, then it is clear that $\alpha(T_A) = |T_A(Q_0)| = |\lambda_1 \cdots \lambda_N|$, where λ_k is the kth diagonal entry. Hence, the assertion is obvious in this case. On the other hand, in the general case, we can find an orthogonal matrix $\mathcal{O}$ such that $A = \mathcal{O}\Lambda\mathcal{O}^T$, where Λ is a diagonal matrix whose diagonal entries are the eigenvalues of A. Hence, by Steps 5) and 6), $\alpha(A) = \alpha(\mathcal{O})\alpha(\Lambda)\alpha(\mathcal{O}^T) = \alpha(\Lambda) = |\det(A)|$.

Step 8). *For every non-singular matrix A, $\alpha(T_A) = |\det(A)|$.*

Set $B = (AA^T)^{1/2}$. Then B is symmetric and $\det(B) = |\det(A)|$. Next set $\mathcal{O} = B^{-1}A$ and note that $\mathcal{O}^T = A^T B^{-1}$ and so $\mathcal{O}\mathcal{O}^T = B^{-1}AA^T B^{-1} = B^{-1}B^2 B^{-1} = \mathbf{I}_{\mathbb{R}^N}$, where $\mathbf{I}_{\mathbb{R}^N}$ denotes the identity matrix. In other words, $\mathcal{O}$ is orthogonal. Since $A = B\mathcal{O}$, we now have that $\alpha(A) = \alpha(B) = |\det(A)|$.

Step 9). *If A is singular, then $\alpha(T_A) = 0$.*

Choose $y \in \mathbb{R}^N$ with unit length so that $y \perp \text{Range}(T_A)$. Next, choose an orthogonal $\mathcal{O}$ so that $\mathbf{e}_1 = T_{\mathcal{O}}(y)$. Then $\mathbf{e}_1 \perp \text{Range}(T_{\mathcal{O}} \circ T_A)$ and so there a rectangle $\hat{I}$ in $\mathbb{R}^{N-1}$ such that $T_{\mathcal{O}} \circ T_A(Q_0) \subseteq \{0\} \times \hat{I}$. But $\{0\} \times \hat{I}$ has measure 0, and therefore $\alpha(T_A) = \alpha(T_{\mathcal{O}} \circ T_A) = 0$. ∎

II.2.3 Exercise.

Show that if H is a **hyperplane** in $\mathbb{R}^N$ (i.e., $H = \{y \in \mathbb{R}^N : y - c \perp \ell\}$ for some $c \in \mathbb{R}^N$ and $\ell \in \mathbb{R}^N \setminus \{0\}$) then $|H| = 0$.

II.2.4 Exercise.

If $\mathbf{v}_1, \cdots,$ and $\mathbf{v}_N$ are vectors in $\mathbb{R}^N$, the **parallelepiped spanned by** $\{\mathbf{v}_1, \cdots, \mathbf{v}_N\}$ is the set

$$P(\mathbf{v}_1, \ldots, \mathbf{v}_N) \equiv \left\{ \sum_1^N x^i \mathbf{v}_i : x^i \in [0,1] \text{ for all } 1 \leq i \leq N \right\}.$$

When $N \geq 2$, the classical prescription for computing the *volume* of a parallelepiped is as the product of *the area of any one side* times the length of the corresponding *altitude*. In analytic terms, this means that the volume is 0 if the vectors $\mathbf{v}_1, \ldots, \mathbf{v}_N$ are linearly dependent and that otherwise the volume of $P(\mathbf{v}_1, \ldots, \mathbf{v}_N)$ can be computed by taking the product of the volume of $P(\mathbf{v}_1, \ldots, \mathbf{v}_{N-1})$, thought of as a subset of the hyperplane $H(\mathbf{v}_1, \ldots, \mathbf{v}_{N-1})$ spanned by $\mathbf{v}_1, \ldots, \mathbf{v}_{N-1}$, times the distance between the vector $\mathbf{v}_N$ and the hyperplane $H(\mathbf{v}_1, \ldots, \mathbf{v}_{N-1})$. Using Theorem II.2.2, show that this prescription is correct when the *volume* of a set is interpreted as the LEBESGUE measure of that set.

Chapter III. Lebesque Integration

III.1. Measure Spaces.

In Chapter II we constructed LEBESGUE's measure on $\mathbb{R}^N$. The result of our efforts was a proof that there is a class $\overline{B}_{\mathbb{R}^N}$ of subsets of $\mathbb{R}^N$ and a map $\Gamma \in \overline{B}_{\mathbb{R}^N} \longmapsto |\Gamma| \in [0, \infty]$ such that: $\overline{B}_{\mathbb{R}^N}$ contains all open sets; $\overline{B}_{\mathbb{R}^N}$ is closed under both complementation and countable unions; $|I| = \text{vol}(I)$ for all rectangles I; and $|\bigcup_1^\infty \Gamma_n| = \sum_1^\infty |\Gamma_n|$ whenever $\{\Gamma_n\}_1^\infty$ is a sequence of mutually disjoint elements of $\overline{B}_{\mathbb{R}^N}$. What we are going to do in this section is discuss a few of the general properties which such structures possess.

Given a set E, we will use $\mathcal{P}(E)$ to denote the **power set** of E; that is $\mathcal{P}(E) \equiv \{\Gamma : \Gamma \subseteq E\}$. An **algebra over** E is an $\mathcal{A} \subseteq \mathcal{P}(E)$ with the properties that

i) $\emptyset \in \mathcal{A}$,

ii) $\Gamma \in \mathcal{A} \implies \Gamma^{\complement} \in \mathcal{A}$,

iii) $\Gamma_1, \Gamma_2 \in \mathcal{A} \implies \Gamma_1 \cup \Gamma_2 \in \mathcal{A}$.

By elementary set-theoretic manipulations, one sees that algebra is also closed under differences as well as finite unions and intersections. A σ-**algebra** over E is an algebra $\mathcal{B}$ which is closed under countable unions. Of course, σ-algebras are also closed under differences as well as countable intersections.

III.1.1 Examples.

i) For any E, $\{\emptyset, E\}$ is the *smallest* algebra over E in the sense that every algebra contains this one.

ii) For any E, $\mathcal{P}(E)$ is the *largest* algebra over E in the sense that it contains every other one.

In fact, both $\{\emptyset, E\}$ and $\mathcal{P}(E)$ are σ-algebras over E. Of course, most of the interesting algebras and σ-algebras lie somewhere in between these two extreme examples. To wit, the σ-algebra $\overline{B}_{\mathbb{R}^N}$ over $\mathbb{R}^N$.

III.1.2 Lemma. *The intersection of any collection of algebras or σ-algebras is again an algebra or σ-algebra. In particular, given any non-empty $\mathcal{C} \subseteq \mathcal{P}(E)$, there is a unique minimal algebra $\mathcal{A}(E; \mathcal{C})$ and a unique minimal σ-algebra $\sigma(E; \mathcal{C})$ over E containing $\mathcal{C}$.*

PROOF: The first assertion is easily checked. Given the first assertion, the second one is handled by considering the collection of all algebras or all σ-algebras over E containing the given $\mathcal{C}$. Noting that neither of these collections can be empty ($\mathcal{P}(E)$ being an element of both), one sees that $\mathcal{A}(E;\mathcal{C})$ and $\sigma(E;\mathcal{C})$ can be constructed by taking intersections. ∎

The σ-algebra $\sigma(E;\mathcal{C})$ is called the σ-algebra **generated by** $\mathcal{C}$. Perhaps the most important examples of σ-algebras which are described in terms of a generating set are those which arise in connection with topological spaces. Namely, if E is a topological space and $\mathfrak{G}$ denotes the class of all open sets in E, then $\mathcal{B}_E \equiv \sigma(E;\mathfrak{G})$ is called the **Borel σ-algebra** or **Borel field** over E , and the elements of $\mathcal{B}_E$ are called the **Borel measurable** subsets of E. (For those who are struck by the similarity between $\mathcal{B}_{\mathbb{R}^N}$ and $\overline{\mathcal{B}}_{\mathbb{R}^N}$, a complete explanation will be forthcoming shortly, in Example III.1.5 below. In the meantime, suffice it to say that, by Theorem II.1.12, $\Gamma \in \overline{\mathcal{B}}_{\mathbb{R}^N}$ if and only if there exist $A, B \in \mathcal{B}_{\mathbb{R}^N}$ such that $A \subseteq \Gamma \subseteq B$ and $|B \setminus A| = 0$.)

Usually the class which generates a σ-algebra is not itself even an algebra. Nonetheless, it often has the property that it is closed under finite intersections. For example, this is the case when the generators are the open sets of some topological space. It is also true of the collection of all rectangles in $\mathbb{R}^N$. In the future, we will call a collection $\mathcal{C} \subseteq \mathcal{P}(E)$ a π-**system** if it is closed under finite intersections. As we will see below, it is useful to know what additional properties a π-system must have in order to be a σ-algebra. For this reason we introduce a notion which complements that of a π-system. Namely, we will say that $\mathcal{H} \subseteq \mathcal{P}(E)$ is a λ-**system over** E if

i) $E \in \mathcal{H}$,

ii) $\Gamma_1, \Gamma_2 \in \mathcal{H}$ and $\Gamma_1 \cap \Gamma_2 = \emptyset \implies \Gamma_1 \cup \Gamma_2 \in \mathcal{H}$,

iii) $\Gamma_1, \Gamma_2 \in \mathcal{H}$ and $\Gamma_1 \subseteq \Gamma_2 \implies \Gamma_2 \setminus \Gamma_1 \in \mathcal{H}$,

iv) $\{\Gamma_n\}_1^\infty \subseteq \mathcal{H}$ and $\Gamma_n \nearrow \Gamma \implies \Gamma \in \mathcal{H}$.

The sense in which λ-systems and π-systems constitute complementary notions is explained in the following useful lemma.

III.1.3 Lemma. *The intersection of an arbitrary collection of π-systems or of λ-systems is again a π-system or a λ-system. Moreover, $\mathcal{B} \subseteq \mathcal{P}(E)$ is a σ-algebra over E if and only if it is both a π-system as well as being a λ-system over E. Finally, if $\mathcal{C} \subseteq \mathcal{P}(E)$ is a π-system, then $\sigma(E;\mathcal{C})$ is the smallest λ-system over E containing $\mathcal{C}$.*

PROOF: The first assertion requires no comment. To prove the second one, it suffices to prove that if $\mathcal{B}$ is both a π-system and a λ-system over E, then it is a σ-algebra over E. To this end, first note that $\emptyset = E \setminus E \in \mathcal{B}$. Second, if $\Gamma_1, \Gamma_2 \in \mathcal{B}$, then $\Gamma_1 \cup \Gamma_2 = \Gamma_1 \cup (\Gamma_2 \setminus \Gamma_3)$ where $\Gamma_3 = \Gamma_1 \cap \Gamma_2$. Hence $\mathcal{B}$ is an algebra over E. Finally, if $\{\Gamma_n\} \subseteq \mathcal{B}$, set $A_n = \bigcup_1^n \Gamma_m$ for $n \geq 1$. Then $\{A_n\}_1^\infty \subseteq \mathcal{B}$ and $A_n \nearrow \bigcup_1^\infty \Gamma_m$. Hence $\bigcup_1^\infty \Gamma_m \in \mathcal{B}$, and so $\mathcal{B}$ is a σ-algebra.

To prove the final assertion, let $\mathcal{C}$ be a π-system and $\mathcal{H}$ the smallest λ-system over E containing $\mathcal{C}$. Clearly $\sigma(E; \mathcal{C}) \supseteq \mathcal{H}$; and so all that we have to do is show that $\mathcal{H}$ is π-system over E. To this end, first set $\mathcal{H}_1 = \{\Gamma \subseteq E : \Gamma \cap \Delta \in \mathcal{H}$ for all $\Delta \in \mathcal{C}\}$. It is then easy to check that $\mathcal{H}_1$ is a λ-system over E. Moreover, since $\mathcal{C}$ is a π-system, $\mathcal{C} \subseteq \mathcal{H}_1$, and therefore $\mathcal{H} \subseteq \mathcal{H}_1$. In other words, $\Gamma \cap \Delta \in \mathcal{H}$ for all $\Gamma \in \mathcal{H}$ and $\Delta \in \mathcal{C}$. Next set $\mathcal{H}_2 = \{\Gamma \subseteq E : \Gamma \cap \Delta \in \mathcal{H}$ for all $\Delta \in \mathcal{H}\}$. Again it is clear that $\mathcal{H}_2$ is a λ-system. Also, by the preceding, $\mathcal{C} \subseteq \mathcal{H}_2$. Hence we have shown that $\mathcal{H}$ is a π-system. ∎

Given a set E and a σ-algebra $\mathcal{B}$ over E, we call the pair $(E, \mathcal{B})$ a **measurable space**. The reason for introducing measurable spaces is that they are the natural place on which to define measures. Namely, if $(E, \mathcal{B})$ is a measurable space, we say that the map $\mu : \mathcal{B} \longrightarrow [0, \infty]$ is a **measure** on $(E, \mathcal{B})$ if $\mu(\emptyset) = 0$ and μ is **countably additive** in the sense that for $\{\Gamma_n\}_1^\infty \subseteq \mathcal{B}$:

$$(\text{III}.1.4) \qquad \mu\left(\bigcup_1^\infty \Gamma_n\right) = \sum_1^\infty \mu(\Gamma_n) \quad \text{if} \quad \Gamma_m \cap \Gamma_n = \emptyset \quad \text{for} \quad m \neq n.$$

When $\mu(E) < \infty$, μ is said to be a **finite measure**, and when $\mu(E) = 1$ it is called a **probability measure**. Given a measurable space $(E, \mathcal{B})$ and a measure μ on $(E, \mathcal{B})$, the triple $(E, \mathcal{B}, \mu)$ is called a **measure space**. The measure space $(E, \mathcal{B}, \mu)$ is said to be a **finite measure space** or a **probability space** according to whether μ is a finite measure or a probability measure on $(E, \mathcal{B})$.

III.1.5 Examples.

i) Our basic examples of measures are those constructed by LEBESGUE. Namely, when $E = \mathbb{R}^N$, $\mathcal{B} = \overline{\mathcal{B}}_{\mathbb{R}^N}$, and $\mu = \lambda_{\mathbb{R}^N}$ where $\lambda_{\mathbb{R}^N}$ is the measure defined by $\lambda_{\mathbb{R}^N}(\Gamma) = |\Gamma|$ for $\Gamma \in \overline{\mathcal{B}}_{\mathbb{R}^N}$.

ii) Given a measure space $(E, \mathcal{B}, \mu)$, one can always extend μ as a measure $\overline{\mu}$ on the σ-algebra $\overline{\mathcal{B}}^\mu$ of sets $\Gamma \subseteq E$ with the property that there exist $A, B \in \mathcal{B}$

such that $A \subseteq \Gamma \subseteq B$ and $\mu(B \setminus A) = 0$; indeed, one simply defines $\overline{\mu}(\Gamma) = \mu(A)$. The σ-algebra $\overline{\mathcal{B}}^{\mu}$ is called the **completion of $\mathcal{B}$ with respect to μ**, and the resulting measure space $(E, \overline{\mathcal{B}}^{\mu}, \overline{\mu})$ is said to be **complete** . In this connection, note that what we have been denoting by $\overline{\mathcal{B}}_{\mathbb{R}^N}$ is the completion of the Borel algebra $\mathcal{B}_{\mathbb{R}^N}$ over $\mathbb{R}^N$ with respect to LEBESGUE's measure $\lambda_{\mathbb{R}^N}$. Thus we really should have been using the hideous notation $\overline{\mathcal{B}}_{\mathbb{R}^N}^{\lambda_{\mathbb{R}^N}}$, but, for obvious reasons of aesthetics, we will continue to reserve $\overline{\mathcal{B}}_{\mathbb{R}^N}$ for the completion of $\mathcal{B}_{\mathbb{R}^N}$ with respect to LEBESGUE's measure.

iii) An easy and useful source of examples of measure spaces are those in which E is a countable set, $\mathcal{B} = \mathcal{P}(E)$, and $\mu(\Gamma) = \sum_{x \in \Gamma} \mu_x$, where $\{\mu_x : x \in E\} \subseteq [0, \infty]$.

iv) As a final example, we point out that measure spaces give rise to other measure spaces by means of restriction. Namely, if $(E, \mathcal{B}, \mu)$ is a measure space and $E' \in \mathcal{B}$ is given, define $\mathcal{B}[E'] = \{\Gamma \cap E' : \Gamma \in \mathcal{B}\}$. Then $\mathcal{B}[E']$ is a σ-algebra over E' and $(E', \mathcal{B}[E'], \mu|_{\mathcal{B}[E']})$ is a measure space.

The following theorem gives some of the basic consequences of (III.1.4).

III.1.6 Theorem. *Let $(E, \mathcal{B}, \mu)$ be a measure space. If $\Gamma_1, \Gamma_2 \in \mathcal{B}$ and $\Gamma_1 \subseteq \Gamma_2$ then $\mu(\Gamma_1) \leq \mu(\Gamma_2)$ and, when $\mu(\Gamma_1) < \infty$, $\mu(\Gamma_2 \setminus \Gamma_1) = \mu(\Gamma_2) - \mu(\Gamma_1)$. Moreover, for $\{\Gamma_n\}_1^{\infty} \subseteq \mathcal{B}$:*

(i)
$$\mu(\Gamma_n) \nearrow \mu(\Gamma) \quad \text{if} \quad \Gamma_n \nearrow \Gamma,$$

(ii)
$$\mu(\Gamma_n) \searrow \mu(\Gamma) \quad \text{if} \quad \Gamma_n \searrow \Gamma \quad \text{and} \quad \mu(\Gamma_1) < \infty,$$

(iii)
$$\mu\left(\bigcup_1^{\infty} \Gamma_n\right) \leq \sum_1^{\infty} \mu(\Gamma_n),$$

and

(iv)
$$\mu\left(\bigcup_1^{\infty} \Gamma_n\right) = \sum_1^{\infty} \mu(\Gamma_n) \quad \text{if} \quad \mu(\Gamma_m \cap \Gamma_n) = 0 \quad \text{for} \quad m \neq n.$$

PROOF: If $\Gamma_1 \subseteq \Gamma_2$, then $\mu(\Gamma_2) = \mu(\Gamma_1) + \mu(\Gamma_2 \setminus \Gamma_1)$, since $\Gamma_2 = \Gamma_1 \cup (\Gamma_2 \setminus \Gamma_1)$. The first assertions follow immediately from this.

To prove i), set $\Gamma_0 = \emptyset$ and define $A_{n+1} = \Gamma_{n+1} \setminus \Gamma_n$ for $n \geq 0$. Then $A_m \cap A_n = \emptyset$ for $m \neq n$, $\Gamma_n = \bigcup_1^n A_m$, and $\Gamma = \bigcup_1^\infty A_n$. Hence

$$\mu(\Gamma_n) = \sum_1^n \mu(A_m) \nearrow \sum_1^\infty \mu(A_m) = \mu(\Gamma).$$

The proof of ii) is accomplished by taking $\Delta_n = \Gamma_1 \setminus \Gamma_n$ and applying the preceding to $\{\Delta_n\}$. (One needs $\mu(\Gamma_1) < \infty$ in order to substact it from both sides.)

To prove iii) and iv), again set $\Gamma_0 = \emptyset$ and take $A_{n+1} = \Gamma_{n+1} \setminus \bigcup_1^n \Gamma_m$ for $n \geq 0$. Then $\Gamma_n = A_n \cup D_n$, where $D_n = \bigcup_{m=1}^{n-1}(\Gamma_n \cap \Gamma_m)$, and $\bigcup_1^\infty \Gamma_n = \bigcup_1^\infty A_n$. Hence, since $A_m \cap A_n = \emptyset$ for $m \neq n$,

$$\mu\left(\bigcup_1^\infty \Gamma_n\right) = \mu\left(\bigcup_1^\infty A_n\right) = \sum_1^\infty \mu(A_n) \leq \sum_1^\infty \mu(\Gamma_n) = \sum_1^\infty (\mu(A_n) + \mu(D_n)).$$

This proves that the inequality in iii) always holds and that the equality in iv) holds when $\mu(D_n) \leq \sum_{m=1}^{n-1} \mu(\Gamma_n \cap \Gamma_m) = 0$ for all $n \geq 2$. ∎

III.1.7 Exercise.

The decomposition of the properties of a σ-algebra in terms of π-systems and λ-systems is not the traditional one. Instead, most of the early books on measure theory used algebras instead of π-systems as the usual source of generating sets. In this case the complementary notion is that of **monotone class**: $\mathcal{M}$ is said to be a monotone class if $\Gamma \in \mathcal{M}$ whenever there exists $\{\Gamma_n\}_1^\infty \subseteq \mathcal{M}$ such that either $\Gamma_n \nearrow \Gamma$ or $\Gamma_n \searrow \Gamma$. Show that $\mathcal{B}$ is a σ-algebra over E if and only if it is both an algebra over E and a monotone class. In addition, show that if $\mathcal{A}$ is an algebra over E, $\sigma(E; \mathcal{A})$ is the smallest monotone class containing $\mathcal{A}$.

III.1.8 Exercise.

Let $(E, \mathcal{B})$ be a measurable space with $\mathcal{B} = \sigma(E; \mathcal{C})$. Suppose that μ and ν are a pair of measures on $(E, \mathcal{B})$ such that $\mu(E) = \nu(E) < \infty$ and $\mu(\Gamma) = \nu(\Gamma)$ for all $\Gamma \in \mathcal{C}$. Assuming that $\mathcal{C}$ is either a π-system or an algebra over E, show that $\mu = \nu$ on $\mathcal{B}$.

III.1.9 Exercise.

Let (E, ρ) be a metric space and suppose that μ is a finite measure on $(E, \mathcal{B}_E)$. Show that the completion $\overline{\mathcal{B}}_E^\mu$ of the BOREL field $\mathcal{B}_E$ with respect to μ coincides with the class of all $\Gamma \subseteq E$ such that for each $\epsilon > 0$ there exists a closed set F and an open set G with the properties that $F \subseteq \Gamma \subseteq G$ and $\mu(G \setminus F) < \epsilon$.

III.1.10 Exercise.

Suppose that $(E_1, \mathcal{B}_1)$ and $(E_2, \mathcal{B}_2)$ are two measurable spaces and that $\Phi : E_1 \longrightarrow E_2$ has the property that $\Phi^{-1}(\Gamma) \in \mathcal{B}_1$ for every element Γ in a collection $\mathcal{C}$ which generates $\mathcal{B}_2$. Show that $\Phi^{-1}(\Gamma) \in \mathcal{B}_1$ for every $\Gamma \in \mathcal{B}_2$. In particular, if E_1 and E_2 are topological spaces and $\mathcal{B}_1$ and $\mathcal{B}_2$ are the corresponding Borel algebras, show that $\Phi^{-1}(\Gamma) \in \mathcal{B}_1$ for every $\Gamma \in \mathcal{B}_2$ if Φ is continuous. Conclude from this that $x + \Gamma \in \mathcal{B}_{\mathbb{R}^N}$ for all $x \in \mathbb{R}^N$ and $\Gamma \in \mathcal{B}_{\mathbb{R}^N}$.

III.1.11 Exercise.

Let μ be a measure on $(\mathbb{R}^N, \mathcal{B}_{\mathbb{R}^N})$ which is translation invariant (i.e., $\mu(x + \Gamma) = \mu(\Gamma)$). In addition, assume that $\mu([0,1]^N) = 1$. Show that $\mu = \lambda_{\mathbb{R}^N}$ on $\mathcal{B}_{\mathbb{R}^N}$.

Hint: First check that $\mu(\partial Q) = 0$ for any cube Q. Second, show that $\mu([0, m\lambda]^N) = m^N \mu([0,\lambda])$ for any $m \in \mathbb{Z}^+$ and $\lambda \in \mathbb{R}$. From these, conclude that $\mu(Q) = |Q|$ for all cubes Q. Finally, deduce the required result.

III.1.12 Exercise.

Given sets Γ_n for $n \geq 1$ define

$$\varlimsup_{n \to \infty} \Gamma_n = \bigcap_{m=1}^{\infty} \bigcup_{n=m}^{\infty} \Gamma_n \quad \text{and} \quad \varliminf_{n \to \infty} \Gamma_n = \bigcup_{m=1}^{\infty} \bigcap_{n=m}^{\infty} \Gamma_n.$$

Observe that

$$\varlimsup_{n \to \infty} \Gamma_n = \left\{ x : x \in \Gamma_n \text{ for infinitely many } n \in \mathbb{Z}^+ \right\}$$

and that

(III.1.13)
$$\varliminf_{n \to \infty} \Gamma_n \subseteq \varlimsup_{n \to \infty} \Gamma_n$$

with equality holding when $\{\Gamma_n\}_1^\infty$ is monotone. One says that $\lim_{n \to \infty} \Gamma_n$ exists if equality holds in (II.1.13), in which case $\lim_{n \to \infty} \Gamma_n \equiv \varliminf_{n \to \infty} \Gamma_n$.

Let $(E, \mathcal{B}, \mu)$ be a measure space and $\{\Gamma_n\}_1^\infty \subseteq \mathcal{B}$. Prove each of the following.

(i)
$$\mu\left(\varliminf_{n\to\infty} \Gamma_n\right) \le \varliminf_{n\to\infty} \mu(\Gamma_n).$$

and

(ii)
$$\varlimsup_{n\to\infty} \mu(\Gamma_n) \le \mu\left(\varlimsup_{n\to\infty} \Gamma_n\right) \quad \text{if} \quad \mu\left(\bigcup_1^\infty \Gamma_n\right) < \infty.$$

In particular, under the condition in ii), conclude that

(iii)
$$\lim_{n\to\infty} \mu(\Gamma_n) = \mu\left(\lim_{n\to\infty} \Gamma_n\right) \quad \text{if} \quad \lim_{n\to\infty} \Gamma_n \text{ exists.}$$

Finally, show that

(iv)
$$\mu\left(\varlimsup_{n\to\infty} \Gamma_n\right) = 0 \quad \text{if} \quad \sum_1^\infty \mu(\Gamma_n) < \infty.$$

The result in **iv**) is often called the **Borel–Cantelli Lemma**, and it has many applications in probability theory.

III.1.14 Exercise.

Let $(E, \mathcal{B}, \mu)$ be a measure space.

i) Assuming that $\Gamma_1, \Gamma_2 \in \mathcal{B}$ and that $\mu(\Gamma_1 \cap \Gamma_2) < \infty$, show that

$$\mu(\Gamma_1 \cup \Gamma_2) = \mu(\Gamma_1) + \mu(\Gamma_2) - \mu(\Gamma_1 \cap \Gamma_2).$$

ii) Let $\{\Gamma_m\}_1^n \subseteq \mathcal{B}$ and assume that $\max_{1 \le m \le n} \mu(\Gamma_m) < \infty$. Show that

$$\mu(\Gamma_1 \cup \cdots \cup \Gamma_n) = -\sum_F (-1)^{\operatorname{card}(F)} \mu(\Gamma_F),$$

where the summation is over non-empty subsets F of $\{1, \ldots, n\}$ and $\Gamma_F \equiv \bigcap_{i \in F} \Gamma_i$.

iii) Although the formula in ii) above is seldom used except in the case when $n = 2$, the following is an interesting application of the general result. Let E be the group of permutations on $\{1, \ldots, n\}$, $\mathcal{B} = \mathcal{P}(E)$, and $\mu(\pi) = \frac{1}{n!}$ for each $\pi \in E$. Denote by A the set of $\pi \in E$ such that $\pi(i) \neq i$ for any $1 \leq i \leq n$. Then one can interpret $\mu(A)$ as the probability that, when the numbers $1, \ldots, n$ are randomly ordered, none of them is placed in the correct position. On the basis of this interpretation, one might suspect that $\mu(A)$ should tend to 0 as $n \to \infty$. However, by direct computation, one can see that this is not the case. Indeed, let Γ_i be the set of $\pi \in E$ such that $\pi(i) = i$. Then $A = \left(\Gamma_1 \cup \cdots \cup \Gamma_n\right)^{\complement}$. Hence,

$$\mu(A) = 1 - \mu(\Gamma_1 \cup \cdots \cup \Gamma_n) = 1 + \sum_F (-1)^{\text{card}(F)} \mu\left(\Gamma_F\right).$$

Show that $\mu(\Gamma_F) = \frac{(n-m)!}{n!}$ if $\text{card}(F) = m$, and conclude from this that $\mu(A) = \sum_0^n \frac{(-1)^m}{m!} \longrightarrow \frac{1}{e}$ as $n \to \infty$.

III.2. Construction of Integrals.

We are now very close to the point at which we can return to the problem of integrating functions on a measure space $(E, \mathcal{B}, \mu)$. Recall, from the introduction to Chapter II, that LEBESGUE's procedure entails the use of sums like

$$\sum_k \frac{k}{2^n} \mu\left(\left\{x \in E : f(x) \in \left[\frac{k}{2^n}, \frac{k+1}{2^n}\right)\right\}\right)$$

to approximate the integral of f on E with respect to μ. In particular, we have to be dealing with functions f for which sets of the form

$$\{f \in \Delta\} \equiv \left\{x \in E : f(x) \in \Delta\right\}$$

are in $\mathcal{B}$ when Δ is an interval; and it is only reasonable that we should call such a function measurable. More generally, given measurable spaces $(E_1, \mathcal{B}_1)$ and $(E_2, \mathcal{B}_2)$, we will say that $\Phi : E_1 \longrightarrow E_2$ is a **measurable map on $(E_1, \mathcal{B}_1)$ into $(E_2, \mathcal{B}_2)$** if

$$\{\Phi \in \Gamma\} \equiv \left\{x \in E_1 : \Phi(x) \in \Gamma\right\} = \Phi^{-1}(\Gamma) \in \mathcal{B}_1 \quad \text{for each} \quad \Gamma \in \mathcal{B}_2.$$

Notice the analogy between the definitions of measurability and continuity. In particular, it is clear that if Φ is a measurable map on $(E_1, \mathcal{B}_1)$ into $(E, \mathcal{B}_2)$ and Ψ is a measurable map on $(E_2, \mathcal{B}_2)$ into $(E_3, \mathcal{B}_3)$, then $\Psi \circ \Phi$ is a measurable map on $(E_1, \mathcal{B}_1)$ into $(E_3, \mathcal{B}_3)$. The following lemma is simply a re-statement of Exercise III.1.10 in the language of measurable functions.

III.2.1 Lemma. *Let $(E_1, \mathcal{B}_1)$ and $(E_2, \mathcal{B}_2)$ be measurable spaces and suppose that $\mathcal{B}_2 = \sigma(E_2; \mathcal{C})$ for some $\mathcal{C} \subseteq \mathcal{P}(E_2)$. If $\Phi : E_1 \longrightarrow E_2$ has the property that $\Phi^{-1}(\Gamma) \in \mathcal{B}_1$ for every $\Gamma \in \mathcal{C}$, then Φ is a measurable map on $(E_1, \mathcal{B}_1)$ into $(E_2, \mathcal{B}_2)$. In particular, if E_1 and E_2 are topological spaces and $\mathcal{B}_i = \mathcal{B}_{E_i}$ for $i \in \{1, 2\}$, then every continuous map on E_1 into E_2 is a measurable map on $(E_1, \mathcal{B}_1)$ into $(E_2, \mathcal{B}_2)$.*

In order to handle certain measurability questions, we introduce at this point a construction to which we will return in Section III.4. Namely, given measurable spaces $(E_1, \mathcal{B}_1)$ and $(E_2, \mathcal{B}_2)$, we define the **product of** $(E_1, \mathcal{B}_1)$ **times** $(E_2, \mathcal{B}_2)$ to be the measurable space $(E_1 \times E_2, \mathcal{B}_1 \times \mathcal{B}_2)$ where

$$\mathcal{B}_1 \times \mathcal{B}_2 \equiv \sigma\Big(E_1 \times E_2; \; \big\{ \Gamma_1 \times \Gamma_2 : \Gamma_1 \in \mathcal{B}_1 \text{ and } \Gamma_2 \in \mathcal{B}_2 \big\} \Big).$$

Also, if Φ_i is a measurable map on $(E_0, \mathcal{B}_0)$ into $(E_i, \mathcal{B}_i)$ for $i \in \{1, 2\}$, then we define the **tensor product of** Φ_1 **times** Φ_2 to be the map $\Phi_1 \otimes \Phi_2 : E_0 \longrightarrow E_1 \times E_2$ given by $\Phi_1 \otimes \Phi_2(x) = \big(\Phi_1(x), \Phi_2(x)\big)$, $x \in E_0$.

III.2.2 Lemma. *Referring to the preceding, suppose that Φ_i is a measurable map on $(E_0, \mathcal{B}_0)$ into $(E_i, \mathcal{B}_i)$ for $i \in \{1, 2\}$. Then $\Phi_1 \otimes \Phi_2$ is a measurable map on $(E_0, \mathcal{B}_0)$ into $(E_1 \times E_2, \mathcal{B}_1 \times \mathcal{B}_2)$. Moreover, if E_1 and E_2 are second countable topological spaces, then $\mathcal{B}_{E_1} \times \mathcal{B}_{E_2} = \mathcal{B}_{E_1 \times E_2}$.*

PROOF: To prove the first assertion, we need only note that if $\Gamma_i \in \mathcal{B}_i$, $i \in \{1, 2\}$, then $\Phi_1 \otimes \Phi_2^{-1}(\Gamma_1 \times \Gamma_2) = \Phi_1^{-1}(\Gamma_1) \cap \Phi_2^{-1}(\Gamma_2) \in \mathcal{B}_0$. As for the second assertion, first note that $G_1 \times G_2$ is open in $E_1 \times E_2$ for every pair of open sets G_1 in E_1 and G_2 in E_2. Hence, even without second countability, $\mathcal{B}_{E_1} \times \mathcal{B}_{E_2} \subseteq \mathcal{B}_{E_1 \times E_2}$. At the same time, with second countability, one can write every open G in $E_1 \times E_2$ as the countable union of sets of the form $G_1 \times G_2$ where G_i is open in E_i. Hence, in this case, the opposite inclusion also holds. ∎

In measure theory one is most interested in real-valued functions. However, for reasons of convenience, it is often handy to allow functions to take values in the **extended real line** $\overline{\mathbb{R}} \equiv [-\infty, \infty]$. Unfortunately, the introduction of $\overline{\mathbb{R}}$ involves some annoying problems. We have already encountered such a problem when we wrote the equation $\mu(\Gamma_2 \setminus \Gamma_1) = \mu(\Gamma_2) - \mu(\Gamma_1)$ in Theorem III.1.6. Our problem there, and the basic one which we want to discuss here, stems from the difficulty of extending the arithmetic operations to include ∞ and $-\infty$. Thus, in an attempt to lay all such technical difficulties to rest once and for all, we will spend a little time discussing them here.

To begin with, we point out that $\overline{\mathbb{R}}$ admits a natural metric with which it becomes compact. Namely, define

$$\overline{\rho}(\alpha, \beta) = \frac{2}{\pi} |\arctan(y) - \arctan(x)|,$$

where $\arctan(\pm\infty) \equiv \pm\frac{\pi}{2}$. Clearly $(\overline{\mathbb{R}}, \overline{\rho})$ is a compact metric space: it is homeomorphic to $[-1, 1]$ under the map $t \in [-1, 1] \longmapsto \tan(\pi t/2)$. Moreover, $\mathbb{R}$, with its usual topology, is imbedded in $\overline{\mathbb{R}}$ as a dense open set. In particular, $\mathcal{B}_{\mathbb{R}} = \mathcal{B}_{\overline{\mathbb{R}}}[\mathbb{R}]$ (cf. **iv)** in Examples III.1.5). Having put a topology and measurable structure on $\overline{\mathbb{R}}$, we will next adopt the following extension to $\overline{\mathbb{R}}$ of multiplication: $(\pm\infty) \cdot 0 = 0 \cdot (\pm\infty) = 0$ and $(\pm\infty) \cdot \alpha = \alpha \cdot (\pm\infty) = \mathrm{sgn}(\alpha)\infty$ if $\alpha \in \overline{\mathbb{R}} \setminus \{0\}$. Although $(\alpha, \beta) \in \overline{\mathbb{R}}^2 \longmapsto \alpha \cdot \beta \in \overline{\mathbb{R}}$ is not continuous, one can easily check that it is a measurable map on $(\overline{\mathbb{R}}^2, \mathcal{B}_{\overline{\mathbb{R}}^2})$ into $(\overline{\mathbb{R}}, \mathcal{B}_{\overline{\mathbb{R}}})$. Unfortunately, the extension of addition presents a knottier problem. Indeed, because we do not know how to interpret $\pm\infty \mp \infty$ in general, we will simply avoid doing so at all by restricting the domain of addition to the set $\widehat{\mathbb{R}^2}$ consisting of $\mathbb{R}^2$ with the two points (∞, ∞) and $(-\infty, -\infty)$ appended. Clearly $\widehat{\mathbb{R}^2}$ is an open subset of $\overline{\mathbb{R}}^2$, and so $\mathcal{B}_{\widehat{\mathbb{R}^2}} = \mathcal{B}_{\overline{\mathbb{R}}^2}[\widehat{\mathbb{R}^2}]$. We define addition $(\alpha, \beta) \in \widehat{\mathbb{R}^2} \longmapsto \alpha + \beta \in \overline{\mathbb{R}}$ so that $(\pm\infty) + \alpha = \alpha + (\pm\infty) = \pm\infty$ if $\alpha \neq \mp\infty$. It is then easy to see that $(\alpha, \beta) \longmapsto \alpha + \beta$ is continuous on $\widehat{\mathbb{R}^2}$ into $\overline{\mathbb{R}}$; and therefore it is certainly a measurable map on $(\widehat{\mathbb{R}^2}, \mathcal{B}_{\widehat{\mathbb{R}^2}})$ into $(\overline{\mathbb{R}}, \mathcal{B}_{\overline{\mathbb{R}}})$. Finally, we complete our discussion of $\overline{\mathbb{R}}$ by pointing out that the lattice operations "$\vee$" and "$\wedge$" both admit unique continuous extensions as maps from $\overline{\mathbb{R}}^2$ into $\overline{\mathbb{R}}$, and therefore are not a source of concern.

Having adopted these conventions, we see that, for any pair of measurable functions f_1 and f_2 on a measurable space $(E, \mathcal{B})$ into $(\overline{\mathbb{R}}, \mathcal{B}_{\overline{\mathbb{R}}})$, Lemma III.2.2 guarantees that the measurability of the $\overline{\mathbb{R}}$-valued maps:

$$x \in E \longmapsto f_1 \cdot f_2(x) \equiv f_1(x) \, f_2(x)$$
$$x \in E \longmapsto (f_1 + f_2)(x) \equiv f_1(x) + f_2(x) \in \overline{\mathbb{R}} \quad \text{if} \quad \mathrm{Range}(f_1 \otimes f_2) \subseteq \widehat{\mathbb{R}^2}$$
$$x \in E \longmapsto f_1 \vee f_2(x) \equiv f_1(x) \vee f_2(x)$$

and

$$x \longrightarrow f_1 \wedge f_2(x) \equiv f_1(x) \wedge f_2(x).$$

Thus, of course, if f is measurable on $(E, \mathcal{B})$ into $(\overline{\mathbb{R}}, \mathcal{B}_{\overline{\mathbb{R}}})$, then so are $f^+ \equiv f \vee 0$, $f^- \equiv -f \wedge 0$, and $|f| = f^+ + f^-$. Finally, from now on we will call measurable maps on $(E, \mathcal{B})$ into $(\overline{\mathbb{R}}, \mathcal{B}_{\overline{\mathbb{R}}})$ **measurable functions on** $(E, \mathcal{B})$.

From the measure-theoretic standpoint, the most elementary functions are those which take on only a finite number of distinct values; thus, we will say that such a function is **simple** . Note that the class of simple functions is closed under the lattice operations "∨" and "∧," multiplication, and, when the sum is defined, under addition. Aside from constant functions, the simplest of the simple functions are those which take their values in $\{0,1\}$. Clearly there is a one-to-one correspondence between $\{0,1\}$-valued functions and subsets of E. Namely, with $\Gamma \subseteq E$ we associate the function χ_Γ defined by

$$\chi_\Gamma(x) = \begin{cases} 1 & \text{if } x \in \Gamma \\ 0 & \text{if } x \notin \Gamma \end{cases}$$

The function χ_Γ is called the **indicator** (or **characteristic**) **function** of the set Γ.

The reason why simple functions play such a central role in measure theory is that their integrals are the easiest to describe. To be precise, let $(E, \mathcal{B}, \mu)$ be a measure space and f a non-negative (i.e., a $[0, \infty]$-valued) measurable function on $(E, \mathcal{B})$ which is simple. We then define the **Lebesgue integral of f on E** to be the number

$$\sum_{\alpha \in \text{Range}(f)} \alpha\mu(f = \alpha),$$

where $\mu(f = \alpha)$ is short-hand for $\mu(\{f = \alpha\})$ which, in turn, is short-hand for $\mu(\{x \in E : f(x) = \alpha\})$. There are various ways in which we will denote the LEBESGUE integral of f, depending on how many details the particular situation demands. The various expressions which we will use are, in decreasing order of information conveyed:

$$\int_E f(x)\,\mu(dx), \quad \int_E f\,d\mu, \quad \text{and} \quad \int f\,d\mu.$$

Further, for $\Gamma \in \mathcal{B}$ we will use

$$\int_\Gamma f(x)\,\mu(dx) \quad \text{or} \quad \int_\Gamma f\,d\mu$$

to denote the LEBESGUE integral of $\chi_\Gamma \cdot f$ on E. Observe that this notation is completely consistent, since we would get precisely the same number by restricting f to Γ and computing the LEBESGUE integral of $f|_\Gamma$ relative to $(\Gamma, \mathcal{B}[\Gamma], \mu|_{\mathcal{B}[\Gamma]})$.

It will turn out that all the basic properties of the LEBESGUE integral rest on consistency results about the definition of the integral. The following lemma is the first such consistency result.

III.2.3 Lemma. *Let* $(E, \mathcal{B}, \mu)$ *be a measure space and* f *a non-negative simple measurable function on* $(E, \mathcal{B})$. *If* $f = \sum_{\ell=1}^{n} \beta_\ell \chi_{\Delta_\ell}$ *where* $\{\beta_1, \ldots, \beta_n\} \subseteq [0, \infty]$ *and* $\{\Delta_1, \ldots, \Delta_n\} \subseteq \mathcal{B}$, *then* $\int f \, d\mu = \sum_{\ell=1}^{n} \beta_\ell \mu(\Delta_\ell)$.

PROOF: Let $\{\alpha_1, \ldots, \alpha_m\}$ denote the distinct values of f and set $\Gamma_k = \{f = \alpha_k\}$ for $1 \leq k \leq m$. Since $\Gamma_k \cap \Gamma_{k'} = \emptyset$ for $k \neq k'$,

$$\sum_{\ell=1}^{n} \beta_\ell \mu(\Delta_\ell) = \sum_{\ell=1}^{n} \beta_\ell \sum_{k=1}^{m} \mu(\Delta_\ell \cap \Gamma_k) = \sum_{k=1}^{m} \sum_{\ell=1}^{n} \beta_\ell \mu(\Delta_\ell \cap \Gamma_k);$$

and so all that we need to show is that $\sum_{\ell=1}^{n} \beta_\ell \mu(\Delta_\ell \cap \Gamma_k) = \alpha_k \mu(\Gamma_k)$ for each $1 \leq k \leq m$. Since $\sum_{\ell=1}^{n} \beta_\ell \chi_{\Delta_\ell \cap \Gamma_k} = \alpha_k \chi_{\Gamma_k}$ for each $1 \leq k \leq m$, we now see that it suffices to treat the case when $f = \alpha \chi_\Gamma$ for some $\alpha \in [0, \infty]$ and $\Gamma \in \mathcal{B}$. Further, it is clear that we may assume that $\alpha \neq 0$, since the only way in which one could have $\sum_{\ell=1}^{n} \beta_\ell \chi_{\Delta_\ell} \equiv 0$ is if $\beta_\ell = 0$ whenever $\Delta_\ell \neq \emptyset$. In other words, what we still have to show is that, for any $\alpha(0, \infty]$ and $\Gamma \in \mathcal{B}$,

$$\sum_{\ell=1}^{n} \beta_\ell \mu(\Delta_\ell) = \alpha \chi_\Gamma \quad \text{if} \quad \sum_{\ell=1}^{n} \beta_\ell \chi_{\Delta_\ell} = \alpha \chi_\Gamma.$$

Set $\mathcal{I} = (\{0, 1\})^n$ and for $\boldsymbol{\eta} = (\eta_1, \ldots, \eta_n) \in \mathcal{I}$ define $\beta_{\boldsymbol{\eta}} = \sum_{1}^{n} \eta_\ell \beta_\ell$ and $\Delta_{\boldsymbol{\eta}} = \bigcap_{\ell=1}^{n} \Delta_\ell^{(\eta_\ell)}$, where $\Delta^{(1)} \equiv \Delta$ and $\Delta^{(0)} \equiv \Delta^\complement$. Then $\Delta_{\boldsymbol{\eta}} \cap \Delta_{\boldsymbol{\eta}'} = \emptyset$ if $\boldsymbol{\eta} \neq \boldsymbol{\eta}'$, and, because

$$\Delta_\ell = \bigcup_{\{\boldsymbol{\eta} \in \mathcal{I} : \eta_\ell = 1\}} \Delta_{\boldsymbol{\eta}}$$

for each $1 \leq \ell \leq n$,

$$\sum_{\boldsymbol{\eta} \in \mathcal{I}} \beta_{\boldsymbol{\eta}} \chi_{\Delta_{\boldsymbol{\eta}}} = \sum_{\boldsymbol{\eta} \in \mathcal{I}} \sum_{\ell=1}^{n} \eta_\ell \beta_\ell \chi_{\Delta_{\boldsymbol{\eta}}} = \sum_{\ell=1}^{n} \beta_\ell \chi_{\Delta_\ell} = \alpha \chi_\Gamma$$

In particular, $\beta_{\boldsymbol{\eta}} = \alpha$ if $\Delta_{\boldsymbol{\eta}} \cap \Gamma \neq \emptyset$, $\beta_{\boldsymbol{\eta}} = 0$ if $\Delta_{\boldsymbol{\eta}} \cap \Gamma^\complement \neq \emptyset$, and $\Gamma \subset \bigcup_{\boldsymbol{\eta} \in \mathcal{I}} \Delta_{\boldsymbol{\eta}}$. From these it is clear that $\beta_{\boldsymbol{\eta}} \in \{0, \alpha\}$ for every $\boldsymbol{\eta} \in \mathcal{I}$ and that

$$\Gamma = \bigcup \{\Delta_{\boldsymbol{\eta}} : \beta_{\boldsymbol{\eta}} = \alpha\}.$$

Hence,

$$\sum_{\ell=1}^{n} \beta_\ell \mu(\Delta_\ell) = \sum_{\ell=1}^{n} \beta_\ell \sum_{\{\boldsymbol{\eta}: \eta_\ell = 1\}} \mu(\Delta_{\boldsymbol{\eta}}) = \sum_{\boldsymbol{\eta} \in \mathcal{I}} \beta_{\boldsymbol{\eta}} \mu(\Delta_{\boldsymbol{\eta}}) = \alpha \mu(\Gamma). \quad \blacksquare$$

The importance of Lemma III.2.3 is already apparent in the next lemma.

III.2.4 Lemma. *Let f and g be non-negative simple measurable functions on $(E, \mathcal{B}, \mu)$. Then, for any $\alpha, \beta \in [0, \infty]$,*

$$\int (\alpha f + \beta g) \, d\mu = \alpha \int f \, d\mu + \beta \int g \, d\mu.$$

In particular, if $f \leq g$, then $\int f \, d\mu \leq \int g \, d\mu$. In fact, if $f \leq g$ and $\int f \, d\mu < \infty$, then

$$\int (g - f) \, d\mu = \int g \, d\mu - \int f \, d\mu.$$

PROOF: Clearly it suffices to prove the first assertion. But if $\{\alpha_1, \ldots, \alpha_m\}$ and $\{\beta_1, \ldots, \beta_n\}$ are the distinct values of f and g, respectively, then $\alpha f + \beta g = \sum_{k=1}^{m+n} \gamma_k \chi_{\Gamma_k}$ where $\gamma_k = \alpha \alpha_k$ and $\Gamma_k = \{f = \alpha_k\}$ for $1 \leq k \leq m$ and $\gamma_k = \beta \beta_{k-m}$ and $\Gamma_k = \{g = \beta_{k-m}\}$ for $m+1 \leq k \leq m+n$. Hence the required result follows immediately from Lemma III.2.3. ∎

In order to extend our definition of the LEBESGUE integral to arbitrary non-negative measurable functions, we want to use a limit procedure. The idea is to approximate such a function by ones which are simple. For example, if f is a non-negative measurable function on $(E, \mathcal{B}, \mu)$ then we might take

$$\phi_n = \sum_{k=0}^{4^n - 1} \frac{k}{2^n} \chi_{\{f \in [k/2^n, (k+1)/2^n)\}} + 2^n \chi_{\{f \geq 2^n\}}$$

for $n \geq 1$. Then each ϕ_n is a non-negative, measurable, simple function, and $\phi_n \nearrow f$ as $n \to \infty$. In fact, $\phi_n \longrightarrow f$ uniformly in $(\overline{\mathbb{R}}, \overline{\rho})$. Thus it would seem reasonable to define the integral of f as
(III.2.5)

$$\lim_{n \to \infty} \int_E \phi_n \, d\mu = \lim_{n \to \infty} \left[\sum_{k=1}^{4^n - 1} \frac{k}{2^n} \mu \left(\frac{k}{2^n} \leq f < \frac{k+1}{2^n} \right) + 2^n \mu(f \geq 2^n) \right].$$

Indeed, since $\phi_n \leq \phi_{n+1}$, Lemma III.2.4 guarantees that this limit exists. However, before adopting this definition, we must first check that the definition is not too dependent on the choice of the approximating sequence. In fact, at the moment, it is not even clear that this definition would coincide with the one we have already given for simple f's. This brings us to our second consistency result.

III.2.6 Lemma. Let $(E, \mathcal{B}, \mu)$ be a measure space and suppose that $\{\phi_n\}_1^\infty$ and ψ are non-negative measurable simple functions on $(E, \mathcal{B})$. If $\phi_n \leq \phi_{n+1}$ for all $n \geq 1$ and if $\psi \leq \lim_{n\to\infty} \phi_n$, then $\int \psi \, d\mu \leq \lim_{n\to\infty} \int \phi_n \, d\mu$. In partic-ular, for any non-negative, measurable function f and any sequence $\{\psi_n\}_1^\infty$ of non-negative, measurable, simple functions ψ which increase to f as $n \longrightarrow \infty$, $\lim_{n\to\infty} \int \psi_n \, d\mu$ is the same as the limit in (III.2.5).

PROOF: We treat several cases.

Case 1). $\mu(\psi = \infty) > 0$.

Since, for each $M < \infty$,

$$\mu(\phi_n > M) \nearrow \mu\left(\bigcup_{n=1}^\infty \{\phi_n > M\}\right) \geq \mu(\psi > M) \geq \mu(\psi = \infty) = \epsilon$$

for some $\epsilon > 0$,

$$\lim_{n\to\infty} \int \phi_n \, d\mu \geq \lim_{n\to\infty} M\mu(\phi_n > M) \geq M\epsilon$$

for all $M < \infty$. Hence, in this case, $\lim_{n\to\infty} \int \phi_n \, d\mu = \infty = \int \psi \, d\mu$.

Case 2). $\mu(\psi > 0) = \infty$.

Because ψ is simple, there is an $\epsilon > 0$ such that $\psi > \epsilon$ whenever $\psi > 0$. Hence, under the stated condition,

$$\mu(\phi_n > \epsilon) \nearrow \mu\left(\bigcup_{n=1}^\infty \{\phi_n > \epsilon\}\right) \geq \mu(\psi > 0) = \infty.$$

But this means that

$$\lim_{n\to\infty} \int \phi_n \, d\mu \geq \lim_{n\to\infty} \epsilon\mu(\phi_n > \epsilon) = \infty = \int \psi \, d\mu.$$

Case 3). $\mu(\psi = \infty) = 0$ and $\mu(\psi > 0) < \infty$.

Set $\hat{E} = \{0 < \psi < \infty\}$. Under the present conditions, $\mu(\hat{E}) < \infty$, $\int \psi \, d\mu = \int_{\hat{E}} \psi \, d\mu$, and $\int \phi_n \, d\mu = \int_{\hat{E}} \phi_n \, d\mu$ for all $n \geq 1$. Hence, without loss in generality, we will assume that $E = \hat{E}$. But then $\mu(E) < \infty$, and, because ψ is simple, there exist $\epsilon > 0$ and $M < \infty$ such that $\epsilon \leq \psi \leq M$. Now let $0 < \delta < \epsilon$ be given, and define $E_n = \{\phi_n \geq \psi - \delta\}$. Then $E_n \nearrow E$ and so

$$\lim_{n \to \infty} \int \phi_n \, d\mu \geq \lim_{n \to \infty} \int_{E_n} \phi_n \, d\mu \geq \lim_{n \to \infty} \left[\int_{E_n} \psi \, d\mu - \delta\mu(E_n) \right]$$

$$= \lim_{n \to \infty} \left[\int \psi \, d\mu - \int_{E_n^{\complement}} \psi \, d\mu - \delta\mu(E_n) \right]$$

$$\geq \int \psi \, d\mu - M \lim_{n \to \infty} \mu(E_n^{\complement}) - \delta\mu(E) = \int \psi \, d\mu - \delta\mu(E),$$

since $\mu(E) < \infty$ and therefore $\mu(E_n^{\complement}) \searrow 0$. Because this holds for arbitrarily small $\delta > 0$, we get our result upon letting $\delta \searrow 0$. ∎

The Lemma III.2.6 allows us to complete the definition of the LEBESGUE integral for non-negative, measurable functions. Namely, if f on $(E, \mathcal{B}, \mu)$ is a non-negative, measurable function, then we define the **Lebesgue integral of f on E with respect to μ** to be the number in (III.2.5); and we will continue to use the same notation to denote integrals. Not only does Lemma III.2.6 guarantee that this definition is consistent with our earlier one for simple f's, but it also makes it clear that the value of $\int f \, d\mu$ does depend on the particular way in which one chooses to approximate f by a non-decreasing sequence of non-negative, measurable, simple functions. Thus, for example, the following extension of Lemma III.2.4 is clear.

III.2.7 Lemma. *If f and g are non-negative, measurable functions on the measure space $(E, \mathcal{B}, \mu)$, then for every $\alpha, \beta \in [0, \infty]$,*

$$\int (\alpha f + \beta g) \, d\mu = \alpha \int f \, d\mu + \beta \int g \, d\mu.$$

In particular, if $f \leq g$, then $\int f \, d\mu \leq \int g \, d\mu$ and $\int (g - f) \, d\mu = \int g \, d\mu - \int f \, d\mu$ so long as $\int f \, d\mu < \infty$.

Obviously $\int f \, d\mu$ reflects the size of a non-negative measurable f. The result which follows makes this statement somewhat more quantitative.

III.2.8 Theorem. (MARKOV'S INEQUALITY) *If f is a non-negative measurable function on $(E, \mathcal{B}, \mu)$, then*

$$(\mathrm{III.2.9}) \qquad \mu(f \geq \lambda) \leq \frac{1}{\lambda} \int_{\{f \geq \lambda\}} f \, d\mu \leq \frac{1}{\lambda} \int f \, d\mu, \qquad \lambda > 0.$$

In particular, $\int f\,d\mu = 0$ if and only if $\mu(f > 0) = 0$; and $\mu(f = \infty) = 0$ if $\int f\,d\mu < \infty$.

PROOF: To prove (III.2.9), simply note that $\lambda\chi_{\{f\geq\lambda\}} \leq \chi_{\{f\geq\lambda\}}\cdot f \leq f$. Clearly (III.2.9) implies that $\mu(f > 0) = \lim_{\epsilon\searrow 0}\mu(f \geq \epsilon) = 0$ if $\int f\,d\mu = 0$. Similarly, if $M = \int f\,d\mu < \infty$, then $\mu(f \geq \lambda) \leq M/\lambda$ for all $\lambda > 0$; and therefore, $\mu(f = \infty) \leq \lim_{\lambda\to\infty}\mu(f \geq \lambda) = 0$. Finally, if $\mu(f > 0) > 0$, then there exists an $\epsilon > 0$ such that $\mu(f \geq \epsilon) \geq \epsilon$; and so (III.2.9) implies that $\int f\,d\mu \geq \epsilon^2 > 0$. ∎

The final step in the definition of the LEBESGUE integral is to extend the definition to measurable functions which can take both signs. To this end, let f be a measurable function on the measure space $(E, \mathcal{B}, \mu)$. Then both $\int f^+\,d\mu$ and $\int f^-\,d\mu$ are defined; and, if we want our integral to be linear, we can do nothing but define $\int f\,d\mu$ as the difference between these two. However, before doing so, we must make sure that this difference is well-defined. With this consideration in mind, we now say that $\int f\,d\mu$ **exists** if $\int f^+\,d\mu \wedge \int f^-\,d\mu < \infty$, in which case we define

$$\int_E f(x)\,\mu(dx) = \int f\,d\mu = \int f^+\,d\mu - \int f^-\,d\mu$$

to be the **Lebesgue integral of f on E**. Observe that if $\int f\,d\mu$ exists, then so does $\int_\Gamma f\,d\mu$ for every $\Gamma \in \mathcal{B}$, and in fact

$$\int_{\Gamma_1\cup\Gamma_2} f\,d\mu = \int_{\Gamma_1} f\,d\mu + \int_{\Gamma_2} f\,d\mu$$

if Γ_1 and Γ_2 are disjoint elements of $\mathcal{B}$. Also, it is clear that, when $\int f\,d\mu$ exists,

(III.2.10) $$\left|\int f\,d\mu\right| \leq \int |f|\,d\mu.$$

In particular, $\int_\Gamma f\,d\mu = 0$ if $\mu(\Gamma) = 0$. Finally, when $\int f^+\,d\mu \wedge \int f^-\,d\mu = \infty$, we do not even attempt to define $\int f\,d\mu$.

Once again, we need a consistency result before we know for sure that our definition accomplishes what we wanted it to do; in this case, the preservation of linearity.

III.2.11 Lemma. *Let f and g be measurable functions on $(E, \mathcal{B}, \mu)$ for which $\int f\,d\mu$ and $\int g\,d\mu$ exist and $\left(\int f\,d\mu, \int g\,d\mu\right) \in \widehat{\mathbb{R}^2}$. Then $\mu\left(f \otimes g \notin \widehat{\mathbb{R}^2}\right) = 0$, $\int_{\{f\otimes g\in\widehat{\mathbb{R}^2}\}}(f + g)\,d\mu$ exists, and*

$$\int_{\{f\otimes g\in\widehat{\mathbb{R}^2}\}} (f + g)\,d\mu = \int f\,d\mu + \int g\,d\mu.$$

PROOF: Set $\hat{E} = \{x \in E : (f(x), g(x)) \in \widehat{\mathbb{R}^2}\}$.

Note that, under the stated conditions, either

$$\int f^+ \, d\mu \vee \int g^+ \, d\mu < \infty \quad \text{or} \quad \int f^- \, d\mu \vee \int g^- \, d\mu < \infty.$$

For definiteness, we will assume that $\int f^- \, d\mu \vee \int g^- \, d\mu < \infty$. As a consequence:

$$\mu(\hat{E}^{\complement}) \leq \mu(f^- \vee g^- = \infty) \leq \mu(f^- = \infty) + \mu(g^- = \infty) = 0$$

and, because $(a+b)^- \leq a^- + b^-$ for $(a,b) \in \widehat{\mathbb{R}^2}$, $\int_{\hat{E}} (f+g)^- \, d\mu < \infty$. Hence, all that remains is to prove the asserted equality.

Note that

$$\int_{\hat{E}} (f+g)^+ \, d\mu$$

$$= \int_{\hat{E} \cap \{f \wedge g \geq 0\}} (f^+ + g^+) \, d\mu + \int_{\hat{E} \cap \{g < 0, f+g \geq 0\}} (f^+ - g^-) \, d\mu$$

$$+ \int_{\hat{E} \cap \{f < 0, f+g \geq 0\}} (g^+ - f^-) \, d\mu$$

$$= \int_{\hat{E} \cap \{f+g \geq 0\}} f^+ \, d\mu - \int_{\hat{E} \cap \{f+g \geq 0\}} f^- \, d\mu$$

$$+ \int_{\hat{E} \cap \{f+g \geq 0\}} g^+ \, d\mu - \int_{\hat{E} \cap \{f+g \geq 0\}} g^- \, d\mu.$$

Similarly,

$$-\int_{\hat{E}} (f+g)^- \, d\mu = \int_{\hat{E} \cap \{f+g < 0\}} f^+ \, d\mu - \int_{\hat{E} \cap \{f+g < 0\}} f^- \, d\mu$$

$$+ \int_{\hat{E} \cap \{f+g < 0\}} g^+ \, d\mu - \int_{\hat{E} \cap \{f+g < 0\}} g^- \, d\mu.$$

Upon adding these two, we get the required result. ∎

Given a measurable function f on $(E, \mathcal{B}, \mu)$, define

$$\|f\|_{L^1(\mu)} = \int |f|\, d\mu.$$

We say that $f : E \longrightarrow \overline{\mathbb{R}}$ is μ-**integrable** if f is a measurable function on $(E, \mathcal{B})$ and $\|f\|_{L^1(\mu)} < \infty$; and we use $L^1(\mu) = L^1(E, \mathcal{B}, \mu)$ to denote the set of all $\mathbb{R}$-valued μ-integrable functions. Note that, from the integration theoretic standpoint there is no loss in generality to assume that $f \in L^1(\mu)$ is $\mathbb{R}$-valued. Indeed, if f is a μ-integrable function, then $\chi_{\{|f|<\infty\}} \cdot f \in L^1(\mu)$, $\|f - \chi_{\{|f|<\infty\}} \cdot f\|_{L^1(\mu)} = 0$, and so integrals involving f and $\chi_{\{|f|<\infty\}} \cdot f$ are indistinguishable. The main reason for insisting that f's in $L^1(\mu)$ be $\mathbb{R}$-valued is so that we have no problems taking linear combinations of them over $\mathbb{R}$. This simplifies the statement of results like the following.

III.2.12 Lemma. *For any measure space* $(E, \mathcal{B}, \mu)$, $L^1(\mu)$ *is a linear space and*

(III.2.13) $$\|\alpha f + \beta g\|_{L^1(\mu)} \le |\alpha|\, \|f\|_{L^1(\mu)} + |\beta|\, \|g\|_{L^1(\mu)}$$

whenever $\alpha, \beta \in \mathbb{R}$ *and* $f, g \in L^1(\mu)$.

PROOF: Simply note that $|\alpha f + \beta g| \le |\alpha|\, |f| + |\beta|\, |g|$. ∎

III.2.14 Remark.

As an application of the preceding inequality, we have that if $f, g, h \in L^1(\mu)$, then

(III.2.15) $$\|f - h\|_{L^1(\mu)} \le \|f - g\|_{L^1(\mu)} + \|g - h\|_{L^1(\mu)}$$

To see this, take $\alpha = \beta = 1$ and replace f and g by $f - g$ and $g - h$ in (III.2.13). Thus $\|f - g\|_{L^1(\mu)}$ looks like a good candidate to be chosen as a metric on $L^1(\mu)$. On the other hand, although, from the standpoint of integration theory, a measurable f for which $\|f\|_{L^1(\mu)} = 0$ might as well be identically 0, there is, in general, no reason why f need be identically 0 as a function. This fact prevents $\|\cdot\|_{L^1(\mu)}$ from being a completely satisfactory measure of size. To overcome this problem, we introduce an equivalence relation on the space of measurable functions. Namely, we write $f \overset{\mu}{\sim} g$ if $\mu(f \ne g) = 0$ (cf. Exercise III.2.16 below). It is then clear that $\|f\|_{L^1(\mu)} = 0$ if an only if 0 is in the $\overset{\mu}{\sim}$ equivalence class $[f]^{\overset{\mu}{\sim}}$ of f. Moreover, $\alpha f' + \beta g' \overset{\mu}{\sim} \alpha f + \beta g$ for $\alpha, \beta \in \mathbb{R}$ and $\|f' - g'\|_{L^1(\mu)} = \|f - g\|_{L^1(\mu)}$ whenever $f' \overset{\mu}{\sim} f$ and $g' \overset{\mu}{\sim} g$. Hence, linear operations and $\|\cdot\|_{L^1(\mu)}$ are well-defined on the quotient space $L^1(\mu)/\overset{\mu}{\sim}$, and $\| \cdot \|_{L^1(\mu)}$ determines a metric there. It is customary to suppress the distinction between f and its equivalence class $[f]^{\mu}_{\sim}$, and so we will adopt this convention whenever the distinction is unimportant. Hence, we will continue to write $L^1(\mu)$ even when we mean $L^1(\mu)/\overset{\mu}{\sim}$ and will use f instead of $[f]^{\mu}_{\sim}$. In particular, $L^1(\mu)$ becomes is this way a vector space over $\mathbb{R}$ on which $\|f - g\|_{L^1(\mu)}$ is a metric. As we will see in the next section (cf. Corollary III.3.12), this metric space is complete.

III.2.16 Exercise.

Let f be an $\overline{\mathbb{R}}$-valued function on the measurable space $(E, \mathcal{B})$. Show that f is measurable if and only if $\{f > a\} \in \mathcal{B}$ for every $a \in \mathbb{R}$ if and only if $\{f \geq a\} \in \mathcal{B}$ for every $a \in \mathbb{R}$. At the same time, check that ">" and "$\leq$" can be replaced by "<" and "$\geq$", respectively; and show that one can restrict ones attention to a's from a dense subset of $\mathbb{R}$. Finally, if g is a second $\overline{\mathbb{R}}$-valued measurable function on $(E, \mathcal{B})$, show that each of the sets $\{f < g\}$, $\{f \leq g\}$, $\{f = g\}$, and $\{f \neq g\}$ is an element of $\mathcal{B}$.

III.3. Convergence of Integrals.

One of the distinct advantages that LEBESGUE's theory of integration has over RIEMANN's approach is that LEBESGUE's integral is wonderfully continuous with respect to convergence of integrands. In the present section we will explore some of these continuity properties. We begin by showing that the class of measurable functions is closed under point-wise convergence.

III.3.1 Lemma. *Let $(E, \mathcal{B})$ be a measurable space and $\{f_n\}_1^\infty$ a sequence of measurable functions on $(E, \mathcal{B})$. Then $\sup_{n \geq 1} f_n$, $\inf_{n \geq 1} f_n$, $\overline{\lim}_{n \to \infty} f_n$, and $\underline{\lim}_{n \to \infty} f_n$ are all measurable functions. In particular,*

$$\Delta \equiv \left\{ x \in E : \lim_{n \to \infty} f_n(x) \text{ exists} \right\} \in \mathcal{B},$$

and the function f given by

$$f(x) = \begin{cases} 0 & \text{if } x \notin \Delta \\ \lim_{n \to \infty} f_n(x) & \text{if } x \in \Delta \end{cases}$$

is measurable on $(E, \mathcal{B})$.

PROOF: We first suppose that $\{f_n\}_1^\infty$ is non-decreasing. It is then clear that

$$\left\{ \lim_{n \to \infty} f_n > a \right\} = \bigcup_{n=1}^\infty \{f_n > a\} \in \mathcal{B}, \qquad a \in \mathbb{R},$$

and therefore (cf. Exercise III.2.16) that $\lim_{n \to \infty} f_n$ is measurable. By replacing f_n with $-f_n$, we see that the same conclusion holds in the case when $\{f_n\}_1^\infty$ is non-increasing.

Next, for an arbitrary sequence $\{f_n\}_1^\infty$ of measurable functions,

$$\left\{ f_1 \vee \cdots \vee f_n : n \in \mathbb{Z}^+ \right\}$$

is a non-decreasing sequence of measurable functions. Hence, by the preceding,

$$\sup_{n \geq 1} f_n = \lim_{n \to \infty} \left(f_1 \vee \cdots \vee f_n \right)$$

is measurable; and a similar argument shows that $\inf_{n \geq 1} f_n$ is measurable. Noting that $\inf_{n \geq m} f_n$ does not decrease as m increases, we also see that

$$\varliminf_{n \to \infty} f_n = \lim_{m \to \infty} \inf_{n \geq m} f_n$$

is measurable; and, of course, the same sort of reasoning leads to the measurability of $\varlimsup_{n \to \infty} f_n$.

Finally, since

$$\Delta \equiv \left\{ x \in E : \lim_{n \to \infty} f_n(x) \text{ exists} \right\} = \left\{ \varlimsup_{n \to \infty} f_n = \varliminf_{n \to \infty} f_n \right\},$$

it is an element of $\mathcal{B}$; and from this it is clear that the function f described in the last part of the statement is measurable. ∎

We are now ready to prove the first of three basic continuity theorems about the LEBESGUE integral. In some ways this first one appears the least surprising in that it really only echoes the result obtained in Lemma III.2.6 and is nothing more than the function version of i) in Theorem III.1.6.

III.3.2 Theorem. (THE MONOTONE CONVERGENCE THEOREM) *Let $\{f_n\}_1^\infty$ be a sequence of measurable functions on the measure space $(E, \mathcal{B}, \mu)$. If $f_n \geq 0$ for all $n \geq 1$ and $f_n \nearrow f$, as $n \to \infty$, then $\int f \, d\mu = \lim_{n \to \infty} \int f_n \, d\mu$.*

PROOF: Obviously $\int f \, d\mu \geq \lim_{n \to \infty} \int f_n \, d\mu$. To prove the opposite inequality, for each $m \geq 1$ choose a non-decreasing sequence $\{\phi_{m,n}\}_{n=1}^\infty$ of non-negative measurable simple functions so that $\phi_{m,n} \nearrow f_m$ as $n \to \infty$. Next, define the non-negative, simple, measurable functions $\psi_n = \phi_{1,n} \vee \cdots \vee \phi_{n,n}$ for $n \geq 1$. One then has that:

$$\psi_n \leq \psi_{n+1} \quad \text{and} \quad \psi_{m,n} \leq \psi_n \leq f_n \quad \text{for all} \quad 1 \leq m \leq n;$$

and therefore

$$f_m \leq \lim_{n \to \infty} \psi_n \leq f \quad \text{for each} \quad m \in \mathbb{Z}^+.$$

In particular, $\psi_n \nearrow f$, and therefore

$$\int f \, d\mu = \lim_{n \to \infty} \int \psi_n \, d\mu.$$

At the same time, $\psi_n \leq f_n$ for all $n \in \mathbb{Z}^+$,; and so we now see that

$$\int f \, d\mu \leq \lim_{n \to \infty} \int f_n \, d\mu. \quad \blacksquare$$

Being an inequality instead of an equality, the second of continuity result is often more useful than the other two. It is the function version of **i)** and **ii)** of Exercise III.1.12.

III.3.3 Theorem. (FATOU'S LEMMA) *Let $\{f_n\}_1^\infty$ be a sequence of functions on the measure space $(E, \mathcal{B}, \mu)$. If $f_n \geq 0$ for all $n \geq 1$, then*

$$\int \varliminf_{n \to \infty} f_n \, d\mu \leq \varliminf_{n \to \infty} \int f_n \, d\mu.$$

In particular, if there exits a $[0, \infty)$-valued, μ-integrable function g such that $f_n \leq g$ for all $n \geq 1$, then

$$\int \varlimsup_{n \to \infty} f_n \, d\mu \geq \varlimsup_{n \to \infty} \int f_n \, d\mu.$$

PROOF: Suppose that we knew the first assertion were true. We could then prove the second assertion by simply replacing f_n throughout by $g - f_n$. We will therefore assume that the f_n's are non-negative and check that

$$\int \varliminf_{n \to \infty} f_n \, d\mu \leq \varliminf_{n \to \infty} \int f_n \, d\mu.$$

To this end, set $h_m = \inf_{n \geq m} f_n$. Then $f_m \geq h_m \nearrow \varliminf_{n \to \infty} f_n$ and so, by the Monotone Convergence Theorem,

$$\int \varliminf_{n \to \infty} f_n \, d\mu = \lim_{m \to \infty} \int h_m \, d\mu \leq \varliminf_{m \to \infty} \int f_m \, d\mu. \quad \blacksquare$$

Before stating the third continuity result, we need to introduce a notion which is better suited to measure theory than is that of ordinary point-wise equality. Namely, we will say that an x-dependent statement about quantities on the measure space $(E, \mathcal{B}, \mu)$ holds **μ-almost everywhere** if the set Δ of x for which the statement fails is an element of $\overline{\mathcal{B}}^{\mu}$ which has **μ-measure 0** (i.e., $\mu(\Delta) = 0$). Thus, if $\{f_n\}_1^{\infty}$ is a sequence of measurable functions on the measure space $(E, \mathcal{B}, \mu)$, we will say that $\{f_n\}_1^{\infty}$ **converges μ-almost everywhere** if $\mu(\{x \in E : \lim_{n \to \infty} f_n(x) \text{ does not exist}\}) = 0$. By Lemma III.3.1, we see that if $\{f_n\}_1^{\infty}$ converges μ-almost everywhere, then there is a measurable f such that $f(x) = \lim_{n \to \infty} f_n(x)$ when $\lim_{n \to \infty} f_n(x)$ exists and $f(x) = 0$ otherwise. More generally, we will say that $\{f_n\}_1^{\infty}$ **converges μ-almost everywhere to** f and will write $f_n \longrightarrow f$ (a.e., μ) if $\mu(\{x \in E : f(x) \ne \lim_{n \to \infty} f_n(x)\}) = 0$. Similarly, if f and g are measurable functions, we write $f = g$ (a.e., μ), $f \le g$ (a.e., μ), or $f \ge g$ (a.e., μ) if $\mu(f \ne g) = 0$, $\mu(f > g) = 0$, or $\mu(f < g) = 0$, respectively. Note that $f = g$ (a.e., μ) is the same statement as $f \overset{\mu}{\sim} g$, discussed in Remark III.2.14.

The following can be thought of as the function version of **iii)** of Exercise III.1.12.

III.3.4 Theorem. (LEBESGUE'S DOMINATED CONVERGENCE THEOREM) *Suppose that $\{f_n\}_1^{\infty}$ is a sequence of measurable functions on $(E, \mathcal{B}, \mu)$, and let f be a measurable function to which $\{f_n\}_1^{\infty}$ converges μ-almost everywhere. If there is an μ-integrable function g for which $|f_n| \le g$ (a.e., μ), $n \ge 1$, then f is integrable and $\lim_{n \to \infty} \int |f_n - f| \, d\mu = 0$. In particular, $\int f \, d\mu = \lim_{n \to \infty} \int f_n \, d\mu$.*

PROOF: Let $\hat{E}$ be the set of $x \in E$ for which $f(x) = \lim_{n \to \infty} f_n(x)$ and $\sup_{n \ge 1} |f_n(x)| \le g(x)$. Then $\hat{E}$ is measurable and $\mu(\hat{E}^{\complement}) = 0$, and so integrals over $\hat{E}$ are the same as those as over E. Thus, without loss of generality, we will assume that all the statements hold for every $x \in E$. But then, $f = \lim_{n \to \infty} f_n$, $|f| \le g$ and $|f - f_n| \le 2g$. Hence, by the second part of FATOU's Lemma,

$$\varlimsup_{n \to \infty} \int |f - f_n| \, d\mu \le \int \varlimsup_{n \to \infty} |f - f_n| \, d\mu = 0. \quad \blacksquare$$

In many applications, it is difficult to find the "LEBESGUE dominant" g. For this reason, the following variation of FATOU's Lemma is interesting and often useful.

III.3.5 Theorem. (Lieb's Version of Fatou's Lemma) *Let* $(E, \mathcal{B}, \mu)$ *be a measure space,* $\{f_n\}_1^\infty \cup \{f\} \subseteq L^1(\mu)$, *and assume that* $f_n \longrightarrow f$ *(a.e.,* μ). *Then*

(III.3.6)
$$\lim_{n \to \infty} \left| \|f_n\|_{L^1(\mu)} - \|f\|_{L^1(\mu)} - \|f_n - f\|_{L^1(\mu)} \right|$$
$$= \lim_{n \to \infty} \int \left| |f_n| - |f| - |f_n - f| \right| d\mu = 0.$$

In particular, if $\|f_n\|_{L^1(\mu)} \longrightarrow \|f\|_{L^1(\mu)} < \infty$, *then* $\|f - f_n\|_{L^1(\mu)} \longrightarrow 0$.

Proof: Since

$$\left| \|f_n\|_{L^1(\mu)} - \|f\|_{L^1(\mu)} - \|f_n - f\|_{L^1(\mu)} \right| \leq \int \left| |f_n| - |f| - |f_n - f| \right| d\mu, \quad n \geq 1,$$

we need only check the second equality in (III.3.6). But, because

$$\left| |f_n| - |f| - |f_n - f| \right| \longrightarrow 0 \quad \text{(a.e., } \mu)$$

and

$$\left| |f_n| - |f| - |f_n - f| \right| \leq \left| |f_n| - |f_n - f| \right| + |f| \leq 2|f|,$$

(III.3.6) follows from Lebesgue's Dominated Convergence Theorem. ∎

We now have a great deal of evidence that almost everywhere convergence of integrands often leads to convergence of the corresponding integrals. We next want to see what can be said about the converse implication. To begin with, we point out that $\|f_n\|_{L^1(\mu)} \longrightarrow 0$ **does not** imply that $f_n \longrightarrow 0$ (a.e., μ). Indeed, define the functions $\{f_\ell\}_1^\infty$ on $[0,1]$ so that, for $n \geq 1$ and $2^{n-1} \leq \ell < 2^n$,

$$f_\ell(x) = \begin{cases} 1 & \text{if } 2^n x \in [2(\ell - 2^{n-1}), 2(\ell + 1 - 2^{n-1})] \\ 0 & \text{otherwise.} \end{cases}$$

It is then clear that the f_ℓ's are non-negative and measurable on $([0,1], \mathcal{B}_{[0,1]})$ and that $\overline{\lim}_{\ell \to \infty} f_\ell(x) = 1$ for every $x \in [0,1]$. On the other hand,

$$\int_{[0,1]} f_\ell(x)\, dx = 1/2^n \quad \text{if } 2^{n-1} \leq \ell < 2^n,$$

and therefore $\int_{[0,1]} f_\ell(x)\, dx \longrightarrow 0$ as $\ell \to \infty$.

The preceding discussion makes it clear that it may be useful to consider other notions of convergence. Keeping in mind that we are looking for a type of convergence which can be tested using integrals, we take a hint from MARKOV's inequality (cf. Theorem III.2.8) and say that the sequence $\{f_n\}_1^\infty$ of measurable functions on the measure space $(E, \mathcal{B}, \mu)$ **converges in μ-measure** to the measurable function f if $\mu(|f_n - f| \geq \epsilon) \longrightarrow 0$ as $n \to \infty$ for every $\epsilon > 0$; in which case we will write $f_n \longrightarrow f$ in μ-**measure**. Note that, by MARKOV's inequality (III.2.9), if $\|f_n - f\|_{L^1(\mu)} \longrightarrow 0$ then $f_n \longrightarrow f$ in μ-measure. Hence, this sort of convergence can be easily tested with integrals (cf. Exercise III.3.20 below); and, as a consequence, we see that convergence in μ-measure certainly does not imply convergence μ-almost everywhere. In fact, it takes a moment to see in what sense the limit is even uniquely determined by convergence in μ-measure. For this reason, suppose that $\{f_n\}_1^\infty$ converges to both f and to g in μ-measure. Then, for $\epsilon > 0$,

$$\mu(|f - g| \geq \epsilon) \leq \mu(|f - f_n| \geq \epsilon/2) + \mu(|f_n - g| \geq \epsilon/2) \longrightarrow 0.$$

Hence,

$$\mu(f \neq g) = \lim_{\epsilon \searrow 0} \mu(|f - g| \geq \epsilon) = 0,$$

and so $f = g$ (a.e., μ). That is, convergence in μ-measure determines the limit function to precisely the same extent as either μ-almost everywhere or $\|\cdot\|_{L^1(\mu)}$-convergence does. In particular, from the standpoint of μ-integration theory, convergence in μ-measure has unique limits.

The following theorem should help to explain the relationship between convergence in μ-measure and μ-almost everywhere convergence.

III.3.7 Theorem. *Let $\{f_n\}_1^\infty$ be a sequence of $\mathbb{R}$-valued measurable functions on the measure space $(E, \mathcal{B}, \mu)$.*

i) *There is a measurable function f for which*

$$\text{(III.3.8)} \qquad \lim_{m \to \infty} \mu\left(\sup_{n \geq m} |f - f_n| \geq \epsilon\right) = 0, \qquad \epsilon > 0,$$

if and only if

$$\text{(III.3.9)} \qquad \lim_{m \to \infty} \mu\left(\sup_{n \geq m} |f_n - f_m| \geq \epsilon\right) = 0, \qquad \epsilon > 0.$$

Moreover, (III.3.8) implies that $f_n \longrightarrow f$ both (a.e., μ) and in μ-measure.

ii) *There is a measurable function f to which $\{f_n\}_1^\infty$ converges in μ-measure if and only if*

(III.3.10) $$\lim_{m \to \infty} \sup_{n \geq m} \mu(|f_n - f_m| \geq \epsilon) = 0, \qquad \epsilon > 0.$$

Furthermore, if $f_n \longrightarrow f$ in μ-measure, then there is a subsequence $\{f_{n_j}\}_{j=1}^\infty$ with the property that

$$\lim_{i \to \infty} \mu\left(\sup_{j \geq i}|f - f_{n_j}| \geq \epsilon\right) = 0, \qquad \epsilon > 0;$$

and therefore $f_{n_j} \longrightarrow f$ (a.e., μ) as well as in μ-measure.

iii) *When $\mu(E) < \infty$, $f_n \longrightarrow f$ (a.e., μ) implies (III.3.8) and therefore that $f_n \longrightarrow f$ in μ-measure.*

PROOF: Set
$$\Delta = \left\{x \in E \lim_{n \to \infty} f_n(x) \text{ does not exist}\right\}.$$

For $m \geq 1$ and $\epsilon > 0$, define $\Delta_m(\epsilon) = \{\sup_{n \geq m}|f_n - f_m| \geq \epsilon\}$. It is then easy to check (from CAUCHY's convergence criterion for $\mathbb{R}$) that

$$\Delta = \bigcup_{\ell=1}^\infty \bigcap_{m=1}^\infty \Delta_m(1/\ell).$$

Since (III.3.9) implies that $\mu\left(\bigcap_{m=1}^\infty \Delta_m(\epsilon)\right) = 0$ for every $\epsilon > 0$, and, by the preceding,

$$\mu(\Delta) \leq \sum_{\ell=1}^\infty \mu\left(\bigcap_{m=1}^\infty \Delta_m(1/\ell)\right),$$

we see that (III.3.9) does indeed imply that $\{f_n\}_1^\infty$ converges μ-almost everywhere. In addition, if f is a function to which $\{f_n\}_1^\infty$ converges μ-almost everywhere, then

$$\sup_{n \geq m}|f_n - f| \leq \sup_{n \geq m}|f_n - f_m| + |f_m - f| \leq 2\sup_{n \geq m}|f_n - f_m| \quad \text{(a.e., } \mu\text{)};$$

and so (III.3.9) leads to the existence of an f for which (III.3.8) holds.

Next, suppose that (III.3.8) holds for some f. Then it is obvious that $f_n \longrightarrow f$ both (a.e., μ) and in μ-measure. In addition, (III.3.9) follows immediately from

$$\mu\left(\sup_{n \geq m}|f_n - f_m| \geq \epsilon\right) \leq \mu\left(\sup_{n \geq m}|f_n - f| \geq \epsilon/2\right) + \mu\left(\sup_{n \geq m}|f - f_m| \geq \epsilon/2\right).$$

We now turn to part **ii)**. To see that $f_n \longrightarrow f$ in μ-measure implies (III.3.10), simply note that

$$\mu(|f_n - f_m| \geq \epsilon) \leq \mu(|f - f_n| \geq \epsilon/2) + \mu(|f - f_m| \geq \epsilon/2).$$

Conversely, assume that (III.3.10) holds, and choose $1 \leq n_1 < \cdots < n_i < \cdots$ so that

$$\sup_{n \geq n_i} \mu\left(|f_n - f_{n_i}| \geq 1/2^{i+1}\right) \leq 1/2^{i+1}, \qquad i \geq 1.$$

Then

$$\mu\left(\sup_{j \geq i} |f_{n_j} - f_{n_i}| \geq 1/2^i\right) \leq \mu\left(\sup_{j \geq i} |f_{n_{j+1}} - f_{n_j}| \geq 1/2^{j+1}\right)$$

$$\leq \sum_{j=i}^{\infty} \mu(|f_{n_{j+1}} - f_{n_j}| \geq 1/2^{j+1}) \leq 1/2^i.$$

From this it is clear that $\{f_{n_i}\}_{i=1}^{\infty}$ satisfies (III.3.9) and therefore that there is an f for which (III.3.8) holds with $\{f_n\}$ replaced by $\{f_{n_i}\}$. Hence, $f_{n_i} \longrightarrow f$ both μ-almost everywhere and in μ-measure. In particular, when combined with (III.3.10), this means that

$$\mu(|f_m - f| \geq \epsilon) \leq \varlimsup_{i \to \infty} \mu(|f_m - f_{n_i}| \geq \epsilon/2) + \varlimsup_{i \to \infty} \mu(|f_{n_i} - f| \geq \epsilon/2)$$

$$\leq \sup_{n \geq m} \mu(|f_n - f_m| \geq \epsilon/2) \longrightarrow 0$$

as $m \to \infty$; and so $f_n \longrightarrow f$ in μ-measure.

Finally to prove **iii)**, assume that $\mu(E) < \infty$ and that $f_n \longrightarrow f$ (a.e., μ). Then, by ii) of Theorem III.1.6,

$$\lim_{m \to \infty} \mu\left(\sup_{n \geq m} |f_n - f| \geq \epsilon\right) = \mu\left(\bigcap_{m=1}^{\infty} \left\{\sup_{n \geq m} |f_n - f| \geq \epsilon\right\}\right) = 0$$

for every $\epsilon > 0$, and therefore (III.3.8) holds. In particular, this means that $f_n \longrightarrow f$ in μ-measure. ∎

Because it is quite important to remember the relationships between the various sorts of convergence discussed in Theorem III.3.7, we will summarize them as follows:

$$\|f_n - f\|_{L^1(\mu)} \longrightarrow 0 \Longrightarrow f_n \longrightarrow f \text{ in } \mu\text{-measure}$$

$$\Longrightarrow \lim_{i \to \infty} \mu\left(\sup_{j \geq i} |f_{n_j} - f| \geq \epsilon\right) = 0, \; \epsilon > 0, \text{ for some subsequence } \{f_{n_i}\}$$

$$\Longrightarrow f_{n_i} \longrightarrow f \text{ (a.e., } \mu)$$

and

$$\mu(E) < \infty \text{ and } f_n \longrightarrow f \text{ (a.e., } \mu) \implies f_n \longrightarrow f \text{ in } \mu\text{-measure.}$$

Notice that, when $\mu(E) = \infty$, μ-almost everywhere convergence *does not* imply μ-convergence. For example, consider the functions $\chi_{[n,\infty)}$ on $\mathbb{R}$ with LEBESGUE's measure.

We next show that, at least so far as Theorem III.3.3 and Theorem III.3.4 are concerned, convergence in μ-measure is just as good as μ-almost everywhere convergence.

III.3.11 Theorem. *Let f and $\{f_n\}_1^\infty$ all be measurable $\mathbb{R}$-valued functions on the measure space $(E, \mathcal{B}, \mu)$, and assume that $f_n \longrightarrow f$ in μ-measure.*
(FATOU'S LEMMA): *If $f_n \geq 0$ (a.e., μ) for each $n \geq 1$, then $f \geq 0$ (a.e., μ) and*

$$\int f \, d\mu \leq \varliminf_{n\to\infty} \int f_n \, d\mu.$$

(LEBESGUE'S DOMINATED CONVERGENCE THEOREM): *If there is an integrable g on $(E, \mathcal{B}, \mu)$ such that $|f_n| \leq g$ (a.e., μ) for each $n \geq 1$, then f is integrable, $\lim_{n\to\infty} \left\| f_n - f \right\|_{L^1(\mu)} = 0$, and so $\int f_n \, d\mu \longrightarrow \int f \, d\mu$ as $n \to \infty$.*
(LIEB'S VERSION OF FATOU'S LEMMA): *If $\sup_{n \geq 1} \|f_n\|_{L^1(\mu)} < \infty$, then f is integrable and*

$$\lim_{n\to\infty} \left| \|f_n\|_{L^1(\mu)} - \|f\|_{L^1(\mu)} - \|f_n - f\|_{L^1(\mu)} \right| = \lim_{n\to\infty} \left\| |f_n| - |f| - |f_n - f| \right\|_{L^1(\mu)} = 0$$

In particular, $\|f_n - f\|_{L^1(\mu)} \longrightarrow 0$ if $\|f_n\|_{L^1(\mu)} \longrightarrow \|f\|_{L^1(\mu)} \in \mathbb{R}$.

PROOF: Each of these results is obtained from the corresponding result for μ-almost everywhere convergent sequences via the same trick. Thus we will prove the preceding statement of FATOU's Lemma and will leave the proofs of the other assertions to the reader.

Assume that the f_n's are non-negative μ-almost everywhere and that $f_n \longrightarrow f$ in μ-measure. Choose a subsequence $\{f_{n_m}\}$ so that $\lim_{m\to\infty} \int f_{n_m} \, d\mu = \varliminf_{n\to\infty} \int f_n \, d\mu$. Next, choose a subsequence $\{f_{n_{m_i}}\}$ of $\{f_{n_m}\}$ so that $f_{n_{m_i}} \longrightarrow f$ (a.e., μ). Because each of the $f_{n_{m_i}}$ is non-negative (a.e., μ), it is now clear that $f \geq 0$ (a.e., μ). In addition, by restricting all integrals to the set $\hat{E}$ on which the $f_{n_{m_i}}$'s are non-negative and $f_{n_{m_i}} \longrightarrow f$, we can apply Theorem III.3.2 to obtain

$$\int f \, d\mu = \int_{\hat{E}} f \, d\mu \leq \varliminf_{i\to\infty} \int_{\hat{E}} f_{n_{m_i}} \, d\mu = \lim_{m\to\infty} \int f_{n_m} \, d\mu = \varliminf_{n\to\infty} \int f_n \, d\mu. \quad \blacksquare$$

An important dividend of these considerations is the fact that $L^1(\mu)$ is a *complete metric space*. More precisely, we have the following corollary.

III.3.12 Corollary. *Let* $\{f_n\}_1^\infty \subseteq L^1(\mu)$. *If*

$$\lim_{m \to \infty} \sup_{n \geq m} \|f_n - f_m\|_{L^1(\mu)} = 0,$$

then there exists an $f \in L^1(\mu)$ *such that* $\|f_n - f\|_{L^1(\mu)} \longrightarrow 0$. *In other words,* $\left(L^1(\mu), \| \cdot \|_{L^1(\mu)}\right)$ *is a complete metric space.*

PROOF: By MARKOV's inequality, we see that (III.3.10) holds. Hence, we can find a measurable f such that $f_n \longrightarrow f$ in μ-measure; and so, by FATOU's Lemma,

$$\|f - f_m\|_{L^1(\mu)} \leq \varliminf_{n \to \infty} \|f_n - f_m\|_{L^1(\mu)} \leq \sup_{n \geq m} \|f_n - f_m\|_{L^1(\mu)} \longrightarrow 0$$

as $m \to \infty$. Finally, since $\sup_{n \geq 1} \|f_n\|_{L^1(\mu)} < \infty$, we also see that f is μ-integrable and therefore may be assumed to be $\mathbb{R}$-valued. ∎

Before closing this discussion, we want to prove a result which is not only useful but also helps to elucidate the structure of $L^1(\mu)$.

III.3.13 Theorem. *Let* $(E, \mathcal{B}, \mu)$ *be a measure space and assume that* $\mu(E) < \infty$. *Given a* π-*system* $\mathcal{C} \subseteq \mathcal{P}(E)$ *which generates* $\mathcal{B}$, *denote by* $\mathcal{S}$ *the set of functions* $\sum_{m=1}^n \alpha_m \chi_{\Gamma_m}$, *where* $n \in \mathbb{Z}^+$, $\{\alpha_m\}_1^n \subseteq \mathbb{Q}$, *and* $\{\Gamma_m\}_1^n \subseteq \mathcal{C} \cup \{E\}$. *Then* $\mathcal{S}$ *is dense in* $L^1(\mu)$. *In particular, if* $\mathcal{C}$ *is countable, then* $L^1(\mu)$ *is a separable metric space.*

PROOF: Denote by $\overline{\mathcal{S}}$ the closure in $L^1(\mu)$ of $\mathcal{S}$. It is then easy to see that $\overline{\mathcal{S}}$ is a vector space over $\mathbb{R}$. In particular, if $f \in L^1(\mu)$ and both f^+ and f^- are elements of $\overline{\mathcal{S}}$, then $f \in \overline{\mathcal{S}}$. Hence, we need only check that every non-negative $f \in L^1(\mu)$ is in $\overline{\mathcal{S}}$. Since every such f is the limit in $L^1(\mu)$ of simple elements of $L^1(\mu)$ and because $\overline{\mathcal{S}}$ is a vector space, we now see that it suffices for us to check that $\chi_\Gamma \in \overline{\mathcal{S}}$ for every $\Gamma \in \mathcal{B}$. But it is easy to see that the class of $\Gamma \subseteq E$ for which $\chi_\Gamma \in \overline{\mathcal{S}}$ is a λ-system over E, and, by hypothesis, it contains the π-system $\mathcal{C}$. Now apply Lemma III.1.3. ∎

III.3.14 Corollary. *Let* (E, ρ) *be a metric space and suppose that* μ *is a measure on* $(E, \mathcal{B}_E)$ *with the property that there exists a non-decreasing sequence of open sets* E_n *such that* $\mu(E_n) < \infty$ *for each* $n \geq 1$ *and* $E = \bigcup_1^\infty E_n$. *For each* $n \in \mathbb{Z}^+$, *denote by* $\mathcal{K}_n$ *the set of bounded,* ρ-*uniformly continuous* functions ϕ which vanish identically off of E_n, and set $\mathcal{K} = \bigcup_{n \in \mathbb{Z}^+} \mathcal{K}_n$. Then $\mathcal{K}$ is dense in $L^1(\mu)$.

PROOF: We will show first that, for each $n \in \mathbb{Z}^+$,

$$\widetilde{\mathcal{K}}_n \equiv \{\phi|_{E_n} : \phi \in \mathcal{K}_n\}$$

is dense in $L^1(E_n, \mathcal{B}_{E_n}, \mu_n)$, where μ_n denotes the restriction of μ to $\mathcal{B}_{E_n}$. In view of Theorem III.3.13, this will follow as soon as we show that χ_G is in the $\| \cdot \|_{L^1(\mu_n)}$-closure of $\widetilde{\mathcal{K}}_n$ for each open $G \subseteq E_n$. If $G = E_n = E$, then there is no problem, since in that case $\chi_G \equiv 1$ is both ρ-uniformly continuous and μ_n-integrable. On the other hand, if $G \subseteq E_n$ is open and $G \neq E$, set $F = G^{\complement}$ and define

$$\phi_m(x) = 1 - \big(1 + \rho(x, F)\big)^{-m}, \quad m \geq 1,$$

where $\rho(x, F) \equiv \inf\{\rho(x, y) : y \in F\}$. Since $\rho(\cdot, F)$ is ρ-uniformly continuous, we see that ϕ_m is ρ-uniformly continuous. In addition, it is easy to check that $0 \leq \phi_m \nearrow \chi_G$ as $m \to \infty$. Hence, by the Monotone Convergence Theorem, it follows that $\phi_m\big|_{E_n} \longrightarrow \chi_G\big|_{E_n}$ in $L^1(E_n, \mathcal{B}_{E_n}, \mu_n)$ as $m \to \infty$.

By the preceding, we now know that if $f \in L^1(\mu)$ vanishes identically off of E_n for some $n \geq 1$, then there is a sequence $\{\phi_m\}_1^\infty \subseteq \mathcal{K}_n$ such that $\|\phi_m - f\|_{L^1(\mu)} \longrightarrow 0$ as $m \to \infty$. At the same time, it is clear, by LEBESGUE's Dominated Convergence Theorem, that for any $f \in L^1(\mu)$, $\|f_n - f\|_{L^1(\mu)} \longrightarrow 0$ as $n \to \infty$, where $f_n \equiv \chi_{E_n} \cdot f$. ∎

III.3.15 Exercise.

Let f be a non-negative, integrable function on the measure space $(E, \mathcal{B}, \nu)$, and define $\mu(\Gamma) = \int_\Gamma f \, d\nu$ for $\Gamma \in \mathcal{B}$. Show that μ is a finite measure on $(E, \mathcal{B})$. In addition, show that μ is **absolutely continuous with respect to** ν in the sense that for each $\epsilon > 0$ there is a $\delta > 0$ with the property that $\mu(\Gamma) < \epsilon$ whenever $\nu(\Gamma) < \delta$.

III.3.16 Exercise.

Let f be a non-negative, measurable function on the measure space $(E, \mathcal{B}, \mu)$. If f is integrable, show that

$$(\text{III.3.17}) \qquad\qquad \lim_{\lambda \to \infty} \lambda \mu(f \geq \lambda) = 0.$$

Next, produce a non-negative measurable f on $([0,1], \mathcal{B}_{[0,1]}, \lambda_{[0,1]})$ ($\lambda_{[0,1]}$ is used here to denote the restriction of $\lambda_{\mathbb{R}}$ to $\mathcal{B}_{[0,1]}$) such that (III.3.17) holds but f fails to be integrable. Finally, show that if f is a non-negative measurable function on the finite measure space $(E, \mathcal{B}, \mu)$, then f is integrable if and only if

$$\sum_{n=1}^{\infty} \mu(f > n) < \infty.$$

III.3.18 Exercise.

Let $(E, \mathcal{B}, \mu)$ be a measure space and let $\{f_n\}_1^{\infty}$ be a sequence of measurable functions on $(E, \mathcal{B})$. Next, suppose that $\{g_n\}_1^{\infty} \subseteq L^1(\mu)$ and that $g_n \longrightarrow g \in L^1(\mu)$ in $L^1(\mu)$. The following variants of FATOU's Lemma and LEBESGUE's Dominated Convergence Theorem are often useful.

i) If $f_n \leq g_n$ (a.e., μ) for each $n \geq 1$, show that

$$\varlimsup_{n \to \infty} \int f_n \, d\mu \leq \int \varlimsup_{n \to \infty} f_n \, d\mu.$$

ii) If $f_n \longrightarrow f$ either in μ-measure or μ-almost everywhere and if $|f_n| \leq g_n$ (a.e., μ) for each $n \geq 1$, show that $\|f_n - f\|_{L^1(\mu)} \longrightarrow 0$ and therefore that $\lim_{n \to \infty} \int f_n \, d\mu = \int f \, d\mu$.

III.3.19 Exercise.

Let $(E, \mathcal{B}, \mu)$ be a measure space. A family $\mathcal{K}$ of measurable functions f on $(E, \mathcal{B}, \mu)$ is said to be **uniformly μ-absolutely continuous** if for each $\epsilon > 0$ there is a $\delta > 0$ such that $\sup_{f \in \mathcal{K}} \int_{\Gamma} |f| \, d\mu \leq \epsilon$ whenever $\Gamma \in \mathcal{B}$ and $\mu(\Gamma) < \delta$; and it is said to be **uniformly μ-integrable** if for each $\epsilon > 0$ there is an $R < \infty$ such that $\sup_{f \in \mathcal{K}} \int_{|f| \geq R} |f| \, d\mu \leq \epsilon$.

i) Show that $\mathcal{K}$ is uniformly μ-integrable if it is uniformly μ-absolutely continuous and

$$\sup_{f \in \mathcal{K}} \|f\|_{L^1(\mu)} < \infty.$$

Conversely, suppose that $\mathcal{K}$ is uniformly μ-integrable and show that it is then necessarily uniformly μ-absolutely continuous and, when $\mu(E) < \infty$, that $\sup_{f \in \mathcal{K}} \|f\|_{L^1(\mu)} < \infty$.

ii) If $\sup_{f \in \mathcal{K}} \int |f|^{1+\delta} \, d\mu < \infty$ for some $\delta > 0$, show that $\mathcal{K}$ is uniformly μ-integrable.

iii) Let $\{f_n\}_1^\infty \subseteq L^1(\mu)$ be given. If $f_n \longrightarrow f$ in $L^1(\mu)$, show that $\{f_n\}_1^\infty \cup \{f\}$ is uniformly μ-absolutely continuous and uniformly μ-integrable. Conversely, assuming that $\mu(E) < \infty$, show that $f_n \longrightarrow f$ in $L^1(\mu)$ if $f_n \longrightarrow f$ in μ-measure and $\{f_n\}_1^\infty$ is uniformly μ-integrable.

iv) Assume that $\mu(E) = \infty$. We say that a family $\mathcal{K}$ of measurable functions f on $(E, \mathcal{B}, \mu)$ is **tight** if for each $\epsilon > 0$ there is a $\Gamma \in \mathcal{B}$ such that $\mu(\Gamma) < \infty$ and $\sup_{f \in \mathcal{K}} \int_{\Gamma^\complement} |f| \, d\mu \leq \epsilon$. Assuming that $\mathcal{K}$ is tight, show that $\mathcal{K}$ is uniformly μ-integrable if and only if it is uniformly μ-absolutely continuous and $\sup_{f \in \mathcal{K}} \|f\|_{L^1(\mu)} < \infty$. Next, show that $\mathcal{K}$ is uniformly μ-integrable if $\mathcal{K}$ is tight and $\sup_{f \in \mathcal{K}} \int |f|^{1+\delta} \, d\mu < \infty$ for some $\delta > 0$. Finally, suppose that $\{f_n\} \subseteq L^1(\mu)$ is tight and that $f_n \longrightarrow f$ in μ-measure. Show that $\|f_n - f\|_{L^1(\mu)} \longrightarrow 0$ if and only if $\{f_n\}$ is uniformly μ-integrable.

III.3.20 Exercise.

Let $(E, \mathcal{B}, \mu)$ be a finite measure space. Show that $f_n \longrightarrow f$ in μ-measure if and only if $\int |f_n - f| \wedge 1 \, d\mu \longrightarrow 0$.

III.3.21 Exercise.

Let (E, ρ) be a metric space and $\{E_n\}_1^\infty$ a non-decreasing sequence of open subsets of E such that $E_n \nearrow E$. Let μ and ν be two measures on $(E, \mathcal{B}_E)$ with the properties that $\mu(E_n) \vee \nu(E_n) < \infty$ for every $n \geq 1$ and $\int \phi \, d\mu = \int \phi \, d\nu$ whenever ϕ is a bounded ρ-uniformly continuous ϕ for which there is an $n \geq 1$ such that $\phi \equiv 0$ off of E_n. Show that $\mu = \nu$ on $\mathcal{B}_E$.

III.4. Products of Measures.

Just before Lemma III.2.2, we introduced the product $(E_1 \times E_2, \mathcal{B}_1 \times \mathcal{B}_2)$ of two measurable spaces $(E_1, \mathcal{B}_1)$ and $(E_2, \mathcal{B}_2)$. We now want to show that if μ_i, $i \in \{1, 2\}$, is a measure on $(E_i, \mathcal{B}_i)$, then, under reasonable conditions, there is a unique measure ν on $(E_1 \times E_2, \mathcal{B}_1 \times \mathcal{B}_2)$ with the property that $\nu(\Gamma_1 \times \Gamma_2) = \mu_1(\Gamma_1) \, \mu(\Gamma_2)$ for all $\Gamma_i \in \mathcal{B}_i$.

The key to the construction of ν is found in the following function analog of π- and λ-systems (cf. Lemma III.1.3). Namely, given a space E, we will say that a collection $\mathcal{L}$ of functions $f : E \longrightarrow (-\infty, \infty]$ is a **semi-lattice** if both f^+ and f^- are in $\mathcal{L}$ whenever $f \in \mathcal{L}$. A subcollection $\mathcal{K} \subseteq \mathcal{L}$ will be called an $\mathcal{L}$-**system** if:

i) $1 \in \mathcal{K}$;

ii) if $f, g \in \mathcal{K}$ and $\{f = \infty\} \cap \{g = \infty\} = \emptyset$, then $g - f \in \mathcal{K}$ whenever either $f \leq g$ or $g - f \in \mathcal{L}$;

iii) if $\alpha, \beta \in [0, \infty)$ and $f, g \in \mathcal{K}$, then $\alpha f + \beta g \in \mathcal{K}$;

iv) if $\{f_n\}_1^\infty \subseteq \mathcal{K}$ and $f_n \nearrow f$, then $f \in \mathcal{K}$ whenever f is bounded or $f \in \mathcal{L}$.

The analog of Lemma III.1.3 in this context is the following.

III.4.1 Lemma. *Let $\mathcal{C}$ be a π-system which generates the σ-algebra $\mathcal{B}$ over E, and let $\mathcal{L}$ be a semi-lattice of functions $f : E \longrightarrow (-\infty, \infty]$. If $\mathcal{K}$ is an $\mathcal{L}$-system and $\chi_\Gamma \in \mathcal{K}$ for every $\Gamma \in \mathcal{C}$, then $\mathcal{K}$ contains every $f \in \mathcal{L}$ which is measurable on $(E, \mathcal{B})$.*

PROOF: First note that $\{\Gamma \subseteq E : \chi_\Gamma \in \mathcal{K}\}$ is a λ-system which contains $\mathcal{C}$. Hence, by Lemma III.1.3, $\chi_\Gamma \in \mathcal{K}$ for every $\Gamma \in \mathcal{B}$. Combined with iii) above, this means that $\mathcal{K}$ contains every non-negative measurable simple function on $(E, \mathcal{B})$.

Next, suppose that $f \in \mathcal{L}$ is measurable on $(E, \mathcal{B})$. Then both f^+ and f^- have the same properties, and, by ii) above, it is enough to show that $f^+, f^- \in \mathcal{K}$ in order to know that $f \in \mathcal{K}$. Thus, without loss in generality, we assume that $f \in \mathcal{L}$ is a non-negative measurable function on $(E, \mathcal{B})$. But in that case f is the non-decreasing limit of non-negative measurable simple functions; and so $f \in \mathcal{K}$ by iv). ∎

The power of Lemma III.4.1 to handle questions involving products is already apparent in the following.

III.4.2 Lemma. *Let $(E_1, \mathcal{B}_1)$ and $(E_2, \mathcal{B}_2)$ be measurable spaces and suppose that f is an $\overline{\mathbb{R}}$-valued measurable function on $(E_1 \times E_2, \mathcal{B}_1 \times \mathcal{B}_2)$. Then for each $x_1 \in E_1$ and $x_2 \in E_2$, $f(x_1, \cdot)$ and $f(\cdot, x_2)$ are measurable functions on $(E_2, \mathcal{B}_2)$ and $(E_1, \mathcal{B}_1)$, respectively. Next, suppose that μ_i, $i \in \{1, 2\}$, is a finite measure on $(E_i, \mathcal{B}_i)$. Then for every measurable function f on $(E_1 \times E_2, \mathcal{B}_1 \times \mathcal{B}_2)$ which is either bounded or non-negative, the functions*

$$\int_{E_2} f(\cdot, x_2) \, \mu_2(dx_2) \quad \text{and} \quad \int_{E_1} f(x_1, \cdot) \, \mu_1(dx_1)$$

are measurable on $(E_1, \mathcal{B}_1)$ and $(E_2, \mathcal{B}_2)$, respectively.

PROOF: Clearly it is enough to check all these assertions when f is bounded and non-negative.

Let $\mathcal{L}$ be the collection of all bounded functions on $E_1 \times E_2$, and define $\mathcal{K}$ to be those elements of $\mathcal{L}$ which have all the asserted properties. It is clear that $\chi_{\Gamma_1 \times \Gamma_2} \in \mathcal{K}$ for all $\Gamma_i \in \mathcal{B}_i$. Moreover, it is easy to check that $\mathcal{K}$ is an $\mathcal{L}$-system. Hence, by Lemma III.4.1, we are done. ∎

III.4.3 Lemma. *Let $(E_1, \mathcal{B}_1, \mu_1)$ and $(E_2, \mathcal{B}_2, \mu_2)$ be finite measure spaces. Then there exists a unique measure ν on $(E_1 \times E_2, \mathcal{B}_1 \times \mathcal{B}_2)$ such that*

$$\nu(\Gamma_1 \times \Gamma_2) = \mu_1(\Gamma_1)\,\mu_2(\Gamma_2) \quad \text{for all} \quad \Gamma_i \in \mathcal{B}_i.$$

Moreover, for every non-negative function f on $(E_1 \times E_2, \mathcal{B}_1 \times \mathcal{B}_2)$,

(III.4.4)

$$
\begin{aligned}
&\int_{E_1 \times E_2} f(x_1, x_2)\, \nu(dx_1 \times dx_2) \\
&= \int_{E_2} \left(\int_{E_1} f(x_1, x_2)\, \mu_1(dx_1) \right) \mu_2(dx_2) \\
&= \int_{E_1} \left(\int_{E_2} f(x_1, x_2)\, \mu_2(dx_2) \right) \mu_1(dx_1).
\end{aligned}
$$

PROOF: The uniqueness of ν is guaranteed by Exercise III.1.8. To prove the existence of ν, define

$$\nu_{1,2}(\Gamma) = \int_{E_2} \left(\int_{E_1} \chi_\Gamma(x_1, x_2)\, \mu_1(dx_1) \right) \mu_2(dx_2)$$

and

$$\nu_{2,1}(\Gamma) = \int_{E_1} \left(\int_{E_2} \chi_\Gamma(x_1, x_2)\, \mu_2(dx_2) \right) \mu_1(dx_1)$$

for $\Gamma \in \mathcal{B}_1 \times \mathcal{B}_2$. Using the Monotone Convergence Theorem, one sees that both $\nu_{1,2}$ and $\nu_{2,1}$ are finite measures on $(E_1 \times E_2, \mathcal{B}_1 \times \mathcal{B}_2)$. Moreover, by the same sort of argument as was used to prove Lemma III.4.2, for every non-negative measurable function on $(E_1 \times E_2, \mathcal{B}_1 \times \mathcal{B}_2)$:

$$\int f\, d\nu_{1,2} = \int_{E_1} \left(\int_{E_2} f(x_1, x_2)\, \mu_1(dx_1) \right) \mu_2(dx_2)$$

and

$$\int f\, d\nu_{2,1} = \int_{E_2} \left(\int_{E_1} f(x_1, x_2)\, \mu_2(dx_2) \right) \mu_1(dx_1).$$

Finally, since $\nu_{1,2}(\Gamma_1 \times \Gamma_2) = \mu(\Gamma_1)\,\mu(\Gamma_2) = \nu_{2,1}(\Gamma_1 \times \Gamma_2)$ for all $\Gamma_i \in \mathcal{B}_i$, we see that both $\nu_{1,2}$ and $\nu_{2,1}$ fulfill the requirements placed on ν. Hence, not only does ν exist, but it is also equal to both $\nu_{1,2}$ and $\nu_{2,1}$; and so the preceding equalities lead to (III.4.4). ∎

In order to extend the preceding construction to measures which need not be finite, we have to introduce a qualified notion of finiteness. Namely, we will say that the measure μ on $(E, \mathcal{B})$ is σ-**finite** and will call $(E, \mathcal{B}, \mu)$ a σ-**finite measure space** if E can be written as the union of a countable number of sets $\Gamma \in \mathcal{B}$ for each of which $\mu(\Gamma) < \infty$. Thus, for example, $(\mathbb{R}^N, \overline{\mathcal{B}}_{\mathbb{R}^N}, \lambda_{\mathbb{R}^N})$ is a σ-finite measure space.

III.4.5 Theorem. (TONELLI'S THEOREM) *Let $(E_1, \mathcal{B}_1, \mu_1)$ and $(E_2, \mathcal{B}_2, \mu_2)$ be σ-finite measure spaces. Then there is a unique measure ν on $(E_1 \times E_2, \mathcal{B}_1 \times \mathcal{B}_2)$ such that $\nu(\Gamma_1 \times \Gamma_2) = \mu_1(\Gamma_1)\mu(\Gamma_2)$ for all $\Gamma_i \in \mathcal{B}_i$. In addition, for every non-negative measurable function f on $(E_1 \times E_2, \mathcal{B}_1 \times \mathcal{B}_2)$, $\int f(\cdot, x_2)\mu_2(dx_2)$ and $\int f(x_1, \cdot)\mu_1(dx_1)$ are measurable on $(E_1, \mathcal{B}_1)$ and $(E_2, \mathcal{B}_2)$, respectively, and (III.4.4) continues to hold.*

PROOF: Choose sequences $\{E_{i,n}\}_{n=1}^{\infty} \subseteq \mathcal{B}_i$ for $i \in \{1, 2\}$ so that $\mu_i(E_{i,n}) < \infty$ for each $n \geq 1$ and $E_i = \bigcup_{n=1}^{\infty} E_{i,n}$. Without loss in generality, we assume that $E_{i,m} \cap E_{i,n} = \emptyset$ for $m \neq n$. For each $n \in \mathbb{Z}^+$, define $\mu_{i,n}(\Gamma_i) = \mu_i(\Gamma_i \cap E_{i,n})$, $\Gamma_i \in \mathcal{B}_i$; and, for $(m, n) \in \mathbb{Z}^{+^2}$, let $\nu_{(m,n)}$ on $(E_1 \times E_2, \mathcal{B}_1 \times \mathcal{B}_2)$ be the measure constructed from $\mu_{1,m}$ and $\mu_{2,n}$ as in Lemma III.4.3.

Clearly, by Lemma III.4.2, for any non-negative measurable function f on $(E_1 \times E_2, \mathcal{B}_1 \times \mathcal{B}_2)$,

$$\int_{E_2} f(\cdot, x_2)\,\mu_2(dx_2) = \sum_{n=1}^{\infty} \int_{E_{2,n}} f(\cdot, x_2)\,\mu_{2,n}(dx_2)$$

is measurable on $(E_1, \mathcal{B}_1)$; and, similarly, $\int_{E_1} f(x_1, \cdot)\,\mu_1(dx_1)$ is measurable on $(E_2, \mathcal{B}_2)$. Finally, the map $\Gamma \in \mathcal{B}_1 \times \mathcal{B}_2 \longmapsto \sum_{m,n=1}^{\infty} \nu_{(m,n)}(\Gamma)$ defines a measure ν_0 on $(E_1 \times E_2, \mathcal{B}_1 \times \mathcal{B}_2)$, and it is easy to check that ν_0 has all the required properties. At the same time, if ν is any other measure on $(E_1 \times E_2, \mathcal{B}_1 \times \mathcal{B}_2)$ for which $\nu(\Gamma_1 \times \Gamma_2) = \mu_1(\Gamma_1)\mu_2(\Gamma_2)$, $\Gamma_i \in \mathcal{B}_i$, then, by the uniqueness assertion proved in Lemma III.4.3, ν coincides with $\nu_{(m,n)}$ on $\mathcal{B}_1 \times \mathcal{B}_2[E_{1,m} \times E_{2,n}]$ for each $(m, n) \in \mathbb{Z}^{+^2}$ and is therefore equal to ν_0 on $\mathcal{B}_1 \times \mathcal{B}_2$. $\blacksquare$

The measure ν constructed in Theorem III.4.5 is called the **product of μ_1 times μ_2** and is denoted by $\mu_1 \times \mu_2$.

III.4.6 Theorem. (FUBINI'S THEOREM) *Let $(E_1, \mathcal{B}_1, \mu_1)$ and $(E_2, \mathcal{B}_2, \mu_2)$ be σ-finite measure spaces and f a measurable function on $(E_1 \times E_2, \mathcal{B}_1 \times \mathcal{B}_2)$. Then the f is $\mu_1 \times \mu_2$-integrable if and only if*

$$\int_{E_1} \left(\int_{E_2} |f(x_1, x_2)|\,\mu_2(dx_2) \right) \mu_1(dx_1) < \infty$$

if and only if

$$\int_{E_2} \left(\int_{E_1} |f(x_1, x_2)| \, \mu_1(dx_1) \right) \mu_2(dx_2) < \infty.$$

Next, set

$$\Lambda_1 = \left\{ x_1 \in E_1 : \int_{E_2} |f(x_1, x_2)| \, \mu_2(dx_2) < \infty \right\}$$

and

$$\Lambda_2 = \left\{ x_2 \in E_2 : \int_{E_1} |f(x_1, x_2)| \, \mu_1(dx_1) < \infty \right\};$$

and define f_i on E_i, $i \in \{1,2\}$, by

$$f_1(x_1) = \begin{cases} \int_{E_2} f(x_1, x_2) \, \mu_2(dx_2) & \text{if } x_1 \in \Lambda_1 \\ 0 & \text{otherwise} \end{cases}$$

and

$$f_2(x_2) = \begin{cases} \int_{E_1} f(x_1, x_2) \, \mu_1(dx_1) & \text{if } x_2 \in \Lambda_2 \\ 0 & \text{otherwise}. \end{cases}$$

Then f_i is an $\mathbb{R}$-valued, measurable function on $(E_i, \mathcal{B}_i)$. Finally, if f is $\mu_1 \times \mu_2$-integrable, then $\mu_i(\Lambda_i^{\complement}) = 0$, $f_i \in L^1(\mu_i)$, and

(III.4.7) $$\int_{E_i} f_i(x_i) \, \mu_i(dx_i) = \int_{E_1 \times E_2} f(x_1, x_2) \, (\mu_1 \times \mu_2)(dx_1 \times dx_2)$$

for $i \in \{1, 2\}$.

PROOF: The first assertion is an immediate consequence of Theorem III.4.5. Moreover, since $\Lambda_i \in \mathcal{B}_i$, it is easy to check that f_i is an $\mathbb{R}$-valued, measurable function on $(E_i, \mathcal{B}_i)$. Finally, if f is $\mu_1 \times \mu_2$-integrable, then, by the first assertion, $\mu_i(\Lambda_i) = 0$ and $f_i \in L^1(\mu_i)$; and one can prove (III.4.7) by applying Theorem III.4.5 to f^+ and f^-. ∎

III.4.8 Exercise.

Let $(E, \mathcal{B}, \mu)$ be a σ-finite measure space. Given a non-negative measurable function f on $(E, \mathcal{B})$, define

$$\Gamma(f) = \left\{ (x, t) \in E \times [0, \infty) : t \leq f(x) \right\}$$

and

$$\widehat{\Gamma}(f) = \{(x,t) \in E \times [0,\infty) : t < f(x)\};$$

and show that both $\Gamma(f)$ and $\widehat{\Gamma}(f)$ are elements of $\mathcal{B} \times \mathcal{B}_{[0,\infty)}$. (**Hint**: consider the map $(x,t) \in E \times (0,\infty) \longmapsto f(x) - t \in [0,\infty]$.) Next, show that if $\{f_n\}_1^\infty$ is a sequence of non-negative measurable functions and $f_n \nearrow f$, then

$$\Gamma(f) = \bigcap_{m=1}^\infty \bigcup_{n=1}^\infty \Gamma\left(\frac{(m+1)f_n}{m}\right) \quad \text{and} \quad \widehat{\Gamma}(f) = \bigcup_{m=1}^\infty \bigcup_{n=1}^\infty \widehat{\Gamma}\left(\frac{(m-1)f_n}{m}\right).$$

Finally, show that for any non-negative measurable f

(III.4.9) $$\mu \times \lambda_{\mathbb{R}}\big(\widehat{\Gamma}(f)\big) = \int_E f \, d\mu = \mu \times \lambda_{\mathbb{R}}\big(\Gamma(f)\big).$$

(**Hint**: Choose the f_n's in the preceding to be simple.)

Clearly (III.4.9) can be interpreted as the statement that *the integral of a non-negative function is the area under its graph.*

III.4.10 Exercise.

Let $(E_1, \mathcal{B}_1, \mu_1)$ and $(E_2, \mathcal{B}_2, \mu_2)$ be σ-finite measure spaces and assume that, for $i \in \{1,2\}$, $\mathcal{B}_i = \sigma(E_i; \mathcal{C}_i)$ where $\mathcal{C}_i$ is a π-system containing a sequence $\{E_{i,n}\}$ such that $E_i = \bigcup_{n=1}^\infty E_{i,n}$ and $\mu_i(E_{i,n}) < \infty$, $n \geq 1$. Show that if ν is a measure on $(E_1 \times E_2, \mathcal{B}_1 \times \mathcal{B}_2)$ with the property that $\nu(\Gamma_1 \times \Gamma_2) = \mu_1(\Gamma_1)\mu_2(\Gamma_2)$ for all $\Gamma_i \in \mathcal{C}_i$, then $\nu = \mu_1 \times \mu_2$. Use this fact to show that for any $M, N \in \mathbb{Z}^+$

$$\lambda_{\mathbb{R}^{M+N}} = \lambda_{\mathbb{R}^M} \times \lambda_{\mathbb{R}^N}$$

on $\mathcal{B}_{\mathbb{R}^{M+N}} = \mathcal{B}_{\mathbb{R}^M} \times \mathcal{B}_{\mathbb{R}^N}$. (Cf. Lemma III.2.2.)

III.4.11 Exercise.

Let $(E_1, \mathcal{B}_1, \mu_1)$ and $(E_2, \mathcal{B}_2, \mu_2)$ be σ-finite measure spaces. Given $\Gamma \in \mathcal{B}_1 \times \mathcal{B}_2$, define

$$\Gamma_{(1)}(x_2) \equiv \left\{x_1 \in E_1 : (x_1, x_2) \in \Gamma\right\} \in \mathcal{B}_1 \quad \text{for} \quad x_2 \in E_2$$

and

$$\Gamma_{(2)}(x_1) \equiv \left\{x_2 \in E_2 : (x_1, x_2) \in \Gamma\right\} \quad \text{for} \quad x_1 \in E_1.$$

Check both that $\Gamma_{(i)}(x_j) \in \mathcal{B}_i$ for each $x_j \in E_j$ and that $x_j \in E_j \longmapsto \mu_i\big(\Gamma_{(i)}(x_j)\big) \in [0,\infty]$ is measurable on $(E_j, \mathcal{B}_j)$ ($\{i,j\} = \{1,2\}$). Finally, show that $\mu_1 \times \mu_2(\Gamma) = 0$ if and only if $\mu_i\Big(\Gamma_{(i)}(x_j)\Big) = 0$ for μ_j-almost every $x_j \in E_j$; and, conclude that $\mu_1\Big(\Gamma_{(1)}(x_2)\Big) = 0$ for μ_2-almost every $x_2 \in E_2$ if and only if $\mu_2\Big(\Gamma_{(2)}(x_1)\Big) = 0$ for μ_1-almost every $x_1 \in E_1$.

III.4.12 Exercise.

The condition that the measure spaces of which one is taking a product be σ-finite is essential if one wants to carry out the program in this section. To see this, let $E_1 = E_2 = (0,1)$ and $\mathcal{B}_1 = \mathcal{B}_2 = \mathcal{B}_{(0,1)}$. Define μ_1 on $(E_1, \mathcal{B}_1)$ so that $\mu_1(\Gamma)$ is the number of elements in Γ ($\equiv \infty$ if Γ is not a finite set) and show that μ_1 is a measure on $(E_1, \mathcal{B}_1)$. Next, take μ_2 to be LEBESGUE's measure $\lambda_{(0,1)}$ on $(E_2, \mathcal{B}_2)$. Show that there is a set $\Gamma \in \mathcal{B}_1 \times \mathcal{B}_2$ such that

$$\int_{E_2} \chi_\Gamma(x_1, x_2)\,\mu_2(dx_2) = 0 \quad \text{for every} \quad x_1 \in E_1$$

but

$$\int_{E_1} \chi_\Gamma(x_1, x_2)\,\mu_1(dx_1) = 1 \quad \text{for every} \quad x_2 \in E_2.$$

(**Hint**: Try $\Gamma = \{(x, x) : 0 < x < 1\}$.) In particular, there is no way that the second equality in (III.4.4) can be made to hold. Notice that it is really the *uniqueness*, not so much the *existence*, in Lemma III.4.3 which fails.

III.4.13 Exercise.

Let $(E, \mathcal{B})$ be a measurable space. Given $-\infty \leq a < b \leq \infty$ and a function $f : (a,b) \times E \longrightarrow \mathbb{R}$ with the properties that $f(\cdot, x) \in C((a,b))$ for every $x \in E$ and $f(t, \cdot)$ is measurable on $(E, \mathcal{B})$ for every $t \in (a,b)$, show that f is measurable on $\big((a,b) \times E, \mathcal{B}_{(a,b)} \times \mathcal{B}\big)$. Next, suppose that $f(\cdot, x) \in C^1((a,b))$ for each $x \in E$, set $f'(t, x) = \frac{d}{dt}f(t, x)$, $x \in E$, and show that f' is measurable on $\big((a,b) \times E, \mathcal{B}_{(a,b)} \times \mathcal{B}\big)$. Finally, suppose that μ is a measure on $(E, \mathcal{B})$ and that there is a $g \in L^1(\mu)$ such that $|f(t, x)| \vee |f'(t, x)| \leq g(x)$ for all $(t, x) \in (a,b) \times E$. Show not only that $\int_E f(\cdot, x)\,\mu(dx) \in C^1((a,b))$ but also that

$$\frac{d}{dt}\int_E f(t, x)\,\mu(dx) = \int_E f'(t, x)\,\mu(dx).$$

Chapter IV. Changes of Variable

IV.0. Introduction.

We have now developed the basic theory of LEBESGUE integration. However, thus far we have nearly no tools with which to compute the integrals which we have shown to exist. The purpose of the present chapter is to introduce a technique which often makes evaluation, or at least estimation, possible. The technique is that of changing variables. In this introduction, we describe the technique in complete generality. In ensuing sections we will give some examples of its applications.

Let $(E_1, \mathcal{B}_1)$ and $(E_2, \mathcal{B}_2)$ be a pair of measurable spaces. Given a measure μ on $(E_1, \mathcal{B}_1)$ and a measurable map Φ on $(E_1, \mathcal{B}_1)$ into $(E_2, \mathcal{B}_2)$, we define the **pushforward** or **image $\Phi^* \mu = \mu \circ \Phi^{-1}$ of μ under Φ** by $\Phi^* \mu(\Gamma) = \mu(\Phi^{-1}(\Gamma))$ for $\Gamma \in \mathcal{B}_2$. It is an easy matter to check that $\Phi^* \mu$ is a measure on $(E_2, \mathcal{B}_2)$.

IV.0.1 Lemma. *For every non-negative measurable function ϕ on $(E_2, \mathcal{B}_2)$,*

$$(\text{IV}.0.2) \qquad \int_{E_2} \phi \, d(\Phi^* \mu) = \int_{E_1} \phi \circ \Phi \, d\mu.$$

Moreover, $\phi \in L^1(E_2, \mathcal{B}_2, \Phi^* \mu)$ if and only if $\phi \circ \Phi \in L^1(E_1, \mathcal{B}_1, \mu)$, and (IV.0.2) holds for all $\phi \in L^1(E_2, \mathcal{B}_2, \Phi^* \mu)$.

PROOF: Clearly it suffices to prove the first assertion. To this end, note that (IV.0.2) holds by definition when f is the indicator of a set $\Gamma \in \mathcal{B}_2$. Hence, it also holds when f is a non-negative measurable simple function on $(E_2, \mathcal{B}_2)$. Thus, by the Monotone Convergence Theorem, it must hold for all non-negative measurable functions on $(E_2, \mathcal{B}_2)$. ∎

The reader should note that Lemma IV.0.1 is really more a definition than it is a true theorem. It is only when a judicious choice of Φ has been made that one gets anything useful from (IV.0.2).

IV.1. Lebesgue Integrals vs. Riemann Integrals.

Our first important example of a change of variables will relate integrals over an arbitrary measure space to integrals on the real line. Namely, given a measurable $\overline{\mathbb{R}}$-valued function f on a measure space $(E, \mathcal{B}, \mu)$, define the **distribution of f under μ** to be the measure $\mu_f \equiv f^* \mu$ on $(\overline{\mathbb{R}}, \mathcal{B}_{\overline{\mathbb{R}}})$. We then have that for any non-negative measurable ϕ on $(\overline{\mathbb{R}}, \mathcal{B}_{\overline{\mathbb{R}}})$:

(IV.1.1) $$\int_E \phi \circ f(x) \, \mu(dx) = \int_{\overline{\mathbb{R}}} \phi(t) \, \mu_f(dt).$$

The reason why it is often useful to make this change of variables is that the integral on the right hand side of (IV.1.1) can be evaluated as the limit of RIEMANN integrals, to which all the fundamental facts of the calculus are applicable.

In order to see how the right hand side of (IV.1.1) leads us to RIEMANN integrals, we will prove a general fact about the relationship between LEBESGUE and RIEMANN integrals on the line. Perhaps the most interesting feature of this result is that it shows that a complete description of the class of RIEMANN integrable functions in terms of continuity properties defies a totally RIEMANNian solution and requires the LEBESGUE notion of *almost everywhere*.

IV.1.2 Theorem. *Let ν be a finite measure on $\big((a, b], \mathcal{B}_{(a,b]}\big)$, where $-\infty < a < b < \infty$, and set $\psi(t) = \nu((a, t])$ for $t \in [a, b]$ ($\psi(a) = \nu(\emptyset) = 0$). Then, ψ is right-continuous on $[a, b)$ and non-decreasing on $[a, b]$. Furthermore, if ϕ is a bounded function on $[a, b]$, then ϕ is RIEMANN integrable on $[a, b]$ with respect to ψ if and only if ϕ is continuous (a.e., ν) on $(a, b]$; in which case, ϕ is measurable on $\big((a, b], \overline{\mathcal{B}}^{\nu}_{(a,b]}\big)$ and*

(IV.1.3) $$\int_{(a,b]} \phi \, d\overline{\nu} = (R) \int_{[a,b]} \phi(t) \, d\psi(t).$$

PROOF: It will be convenient to think of ν as being defined on $\big([a, b], \mathcal{B}_{[a,b]}\big)$ by $\nu(\Gamma) \equiv \nu(\Gamma \cap (a, b])$ for $\Gamma \in \mathcal{B}_{[a,b]}$. Thus, we will do so.

Obviously ψ is right-continuous and non-decreasing on $[a, b]$. In fact, ψ is continuous at a and at precisely those $t \in (a, b]$ for which $\nu(\{t\}) = 0$.

Assume that ϕ is RIEMANN integrable on $[a, b]$ with respect to ψ. To see that ϕ is continuous (a.e., ν) on $(a, b]$, choose, for each $n \geq 1$, a finite, non-overlapping, exact cover $\mathcal{C}_n$ of $[a, b]$ by intervals I such that $\|\mathcal{C}_n\| < 1/n$ and ψ is continuous at I^- for every $I \in \mathcal{C}_n$. Then $\nu(\Delta) = 0$, where $\Delta = \bigcup_{n=1}^{\infty} \{I^- : I \in \mathcal{C}_n\}$. Given $m \geq 1$, let $\mathcal{C}_{m,n}$ be the set of those $I \in \mathcal{C}_n$ such that $\sup_I \phi - \inf_I \phi \geq 1/m$. It is then easy to check that

$$\{t \in (a, b] \setminus \Delta : \phi \text{ is not continuous at } t\} \subseteq \bigcup_{m=1}^{\infty} \bigcap_{n=1}^{\infty} \bigcup \mathcal{C}_{m,n}.$$

But, by Exercise I.2.26,

$$\nu\left(\bigcap_{n=1}^{\infty}\bigcup \mathcal{C}_{m,n}\right) \leq \lim_{n\to\infty}\sum_{I\in\mathcal{C}_{m,n}}\Delta_I\psi = 0,$$

and therefore $\nu\left(\bigcup_{m=1}^{\infty}\bigcap_{n=1}^{\infty}\bigcup \mathcal{C}_{m,n}\right) = 0$. We have therefore shown that ϕ is continuous (a.e., ν) on $(a, b]$.

Conversely, suppose that ϕ is continuous (a.e., ν) on $(a, b]$. Let $\{\mathcal{C}_n\}_1^{\infty}$ be a sequence of finite, non-overlapping, exact covers of $[a, b]$ by intervals I such that $\|\mathcal{C}_n\| \longrightarrow 0$. For each $n \geq 1$, define $\overline{\phi}_n(t) = \sup_I \phi$ and $\underline{\phi}(t) = \inf_I \phi$ for $t \in I \setminus \{I^-\}$ and $I \in \mathcal{C}_n$. Clearly, both $\overline{\phi}_n$ and $\underline{\phi}_n$ are measurable on $\left((a, b], \mathcal{B}_{(a,b]}\right)$. Moreover,

$$\inf_{(a,b]} \phi \leq \underline{\phi}_n \leq \phi \leq \overline{\phi}_n \leq \sup_{(a,b]} \phi$$

for all $n \geq 1$. Finally, since ϕ is continuous (a.e., ν),

$$\phi = \lim_{n\to\infty}\underline{\phi}_n = \lim_{n\to\infty}\overline{\phi}_n \quad (\text{a.e.}, \nu);$$

and so, not only is $\phi = \underline{\lim}_{n\to\infty}\underline{\phi}_n$ (a.e., ν) and therefore measurable on $\left((a, b], \overline{\mathcal{B}}^{\nu}_{(a,b]}\right)$, but also

$$\lim_{n\to\infty}\int_{(a,b]}\underline{\phi}_n\, d\nu = \int_{(a,b]}\phi\, d\overline{\nu} = \lim_{n\to\infty}\int_{(a,b]}\overline{\phi}_n\, d\nu.$$

In particular, we conclude that

$$\sum_{I\in\mathcal{C}_n}\left(\sup_I \phi - \inf_I \phi\right)\Delta_I\psi = \int_{(a,b]}\left(\overline{\phi}_n - \underline{\phi}_n\right)d\nu \longrightarrow 0$$

as $n \to \infty$. From this it is clear both that ϕ is RIEMANN integrable on $[a, b]$ with respect to ψ and that (IV.1.3) holds. ∎

We are now ready to prove the main result to which this section is devoted.

IV.1.4 Theorem. *Let $(E, \mathcal{B}, \mu)$ be a measure space and f a non-negative, measurable function on $(E, \mathcal{B})$. Then $t \in (0, \infty) \longmapsto \mu(f > t) \in [0, \infty]$ is a right-continuous, non-increasing function. In particular, it is measurable on $\left((0, \infty), \mathcal{B}_{(0,\infty)}\right)$ and has at most a countable number of discontinuities. Next,* assume that $\phi \in C([0, \infty)) \cap C^1((0, \infty))$ *is a non-decreasing function satisfying* $\phi(0) = 0 < \phi(t)$, $t > 0$, *and set* $\phi(\infty) = \lim_{t \to \infty} \phi(t)$. *Then*

$$(\text{IV.1.5}) \qquad \int_E \phi \circ f(x)\, \mu(dx) = \int_{(0,\infty)} \phi'(t)\mu(f > t)\, \lambda_{\mathbb{R}}(dt).$$

Hence, either $\mu(f > \delta) = \infty$ *for some* $\delta > 0$, *in which case both sides of* (IV.1.5) *are infinite, or for each* $0 < \delta < r < \infty$ *the map* $t \in [\delta, r] \longmapsto \phi'(t)\mu(f > t)$ *is* RIEMANN *integrable and*

$$\int_E \phi \circ f(x)\, \mu(dx) = \lim_{\substack{\delta \searrow 0 \\ r \nearrow \infty}} (R) \int_{[\delta,r]} \phi'(t)\mu(f > t)\, dt.$$

PROOF: It is clear that $t \in (0, \infty) \longmapsto \mu(f > t)$ is right-continuous and non-increasing. Hence, if $\delta = \sup\{t \in (0, \infty) : \mu(f > t) = \infty\}$, then $\mu(f > t) = \infty$ for $t \in (0, \delta)$ and $t \in (\delta, \infty) \longmapsto \mu(f > t)$ has at most a countable number of discontinuities. Furthermore, if

$$h_n(t) = \mu\big(f > (k+1)/n\big) \quad \text{for} \quad t \in \big(k/n, (k+1)/n\big], k \geq 0, \text{ and } n \geq 1,$$

then each h_n is clearly measurable on $\big((0, \infty), \mathcal{B}_{(0,\infty)}\big)$ and $h_n(t) \longrightarrow \mu(f > t)$ for each $t \in (0, \infty)$. Hence, $t \in (0, \infty) \longmapsto \mu(f > t)$ is measurable on $\big((0, \infty), \mathcal{B}_{(0,\infty)}\big)$.

We next turn to the proof of (IV.1.5). Since

$$\left(\int_E \phi \circ f\, d\mu \right) \wedge \left(\int_{(0,\infty)} \phi'(t)\mu(f > t)\, \lambda_{\mathbb{R}}(dt) \right) \geq \phi(\delta)\mu(f > \delta),$$

it is clear that both sides of (IV.1.5) are infinite when $\mu(f > \delta) = \infty$ for some $\delta > 0$. Thus we will assume that $\mu(f > \delta) < \infty$ for every $\delta > 0$. Then the restriction of μ_f to $\mathcal{B}_{[\delta,\infty)}$ is a finite measure for every $\delta > 0$. Given $\delta > 0$, set $\psi_\delta(t) = \mu_f((\delta, t])$ for $t \in [\delta, \infty)$ and apply Theorem IV.1.2 and Theorem I.2.7 to see that

$$\int_{\{\delta < f \leq r\}} \phi \circ f\, d\mu = \int_{(\delta,r]} \phi\, d\mu_f = (R) \int_{[\delta,r]} \phi(t)\, d\psi_\delta(t)$$

$$= \phi(r)\psi_\delta(r) - (R) \int_{[\delta,r]} \psi_\delta(t)\phi'(t)\, dt$$

$$= \phi(\delta)\psi_\delta(r) + (R) \int_{[\delta,r]} \big(\psi_\delta(r) - \psi_\delta(t)\big)\phi'(t)\, dt$$

$$= \phi(\delta)\mu(\delta < f \leq r) + (R) \int_{[\delta,r]} \mu(t < f \leq r)\phi'(t)\, dt$$

$$= \phi(\delta)\mu(\delta < f \leq r) + \int_{(\delta,r]} \mu(t < f \leq r)\phi'(t)\, \lambda_{\mathbb{R}}(dt)$$

for each $r \in (\delta, \infty)$. Hence, after simple arithmetic manipulation and an application of the Monotone Convergence Theorem, we get

$$\int_{\{\delta < f < \infty\}} (\phi \circ f - \phi(\delta)) \, d\mu = \int_{(\delta, \infty)} \phi'(t) \mu(t < f < \infty) \, \lambda_{\mathbb{R}}(dt)$$

as $r \nearrow \infty$. At the same time, it is clear that

$$\int_{\{f = \infty\}} (\phi \circ f - \phi(\delta)) \, d\mu = [\phi(\infty) - \phi(\delta)] \mu(f = \infty)$$

$$= \int_{(\delta, \infty)} \mu(f = \infty) \phi'(t) \, \lambda_{\mathbb{R}}(dt);$$

and, after combining these, we now arrive at

$$\int_{\{f > \delta\}} (\phi \circ f(x) - \phi(\delta)) \, \mu(dx) = \int_{(\delta, \infty)} \mu(f > t) \phi'(t) \, \lambda_{\mathbb{R}}(dt).$$

Thus, (IV.1.5) will be proved once we show that

$$\lim_{\delta \searrow 0} \int_{\{f > \delta\}} (\phi \circ f(x) - \phi(\delta)) \, \mu(dx) = \int_E \phi \circ f(x) \, \mu(dx).$$

But if $0 < \delta_1 < \delta_2 < \infty$, then $0 \leq (\phi \circ f - \phi(\delta_2)) \chi_{\{f > \delta_2\}} \leq (\phi \circ f - \phi(\delta_1)) \chi_{\{f > \delta_1\}}$, and so the required convergence follows by the Monotone Convergence Theorem.

Finally, to prove the last part of the theorem when $\mu(f > \delta) < \infty$ for every $\delta > 0$, simply note that

$$\int_{(0, \infty)} \phi'(t) \mu(f > t) \, \lambda_{\mathbb{R}}(dt) = \lim_{\substack{\delta \searrow 0 \\ r \nearrow \infty}} \int_{(\delta, r]} \phi'(t) \mu(f > t) \, \lambda_{\mathbb{R}}(dt),$$

and apply Theorem IV.1.2. ∎

IV.1.6 Exercise.

i) Let ψ be a continuous, non-decreasing function on the compact interval $[a, b]$. Show that

$$(\text{IV.1.7}) \quad (R) \int_{[a,b]} f \circ \psi(s) \, d\psi(s) = (R) \int_{[\psi(a), \psi(b)]} f(t) \, dt, \quad f \in C([a,b]).$$

ii) Suppose that μ is a measure on $(\mathbb{R}, \mathcal{B}_{\mathbb{R}})$ with the properties that $\mu(I) < \infty$ for each compact interval I and $\mu(\{t\}) = 0$ for each $t \in \mathbb{R}$. Let $\Phi : \mathbb{R} \longrightarrow \mathbb{R}$ be a function satisfying $\mu([a, b]) = \Phi(b) - \Phi(a)$ for all $-\infty < a < b < \infty$. Note that Φ is necessarily continuous and non-decreasing, and show that $\Phi^* \mu$ coincides with the restriction of $\lambda_{\mathbb{R}}$ to $\mathcal{B}_{\mathbb{R}}[\Phi(\mathbb{R})]$.

IV.1.8 Exercise.

A particularly important case of Theorem IV.1.4 is when $\phi(t) = t^p$ for some $p \in (0, \infty)$, in which case (IV.1.5) yields

$$(IV.1.9) \qquad \int_E |f(x)|^p \, \mu(dx) = p \int_{(0,\infty)} t^{p-1} \mu(|f| > t) \, \lambda_{\mathbb{R}}(dt).$$

Use (IV.1.9) to show that $|f|^p$ is μ-integrable if and only if

$$\sum_{n=1}^{\infty} \frac{1}{n^{p+1}} \mu(|f| > 1/n) + \sum_{n=1}^{\infty} n^{p-1} \mu(|f| > n) < \infty.$$

Compare this result to the one obtained in the last part of Exercise III.3.16.

IV.2. Polar Coordinates.

From now on, at least whenever the meaning is clear from the context, we will use the notation "dx" instead of the more cumbersome "$\lambda_{\mathbb{R}^N}(dx)$" when doing LEBESGUE integration with respect to LEBESGUE's measure on $\mathbb{R}^N$.

In this section, we examine a change of variables which plays an extremely important role in the evaluation of many LEBESGUE integrals over $\mathbb{R}^N$. Let $\mathbf{S}^{N-1}$ denote the $(N-1)$-**sphere** $\{x \in \mathbb{R}^N : |x| = 1\}$ in $\mathbb{R}^N$, and define $\Phi_N : \mathbb{R}^N \setminus \{0\} \longrightarrow \mathbf{S}^{N-1}$ by $\Phi_N(x) = \frac{x}{|x|}$. Clearly Φ_N is continuous. Next, define the finite measure $\lambda_{\mathbf{S}^{N-1}}$ on $\mathbf{S}^{N-1}$ to be the image under Φ_N of $N \cdot \lambda_{\mathbb{R}^N}$ restricted to $\mathcal{B}_{B(0,1) \setminus \{0\}}$. (Recall that $B(a, r)$ denotes the ball of radius r around a in the metric space under consideration; in this case, the metric space is $\mathbb{R}^N$ with the usual EUCLIDEan metric.) Noting that $\Phi_N(rx) = \Phi_N(x)$ for all $r > 0$ and $x \in \mathbb{R}^N \setminus \{0\}$, we conclude from Theorem II.2.2 that $\int_{B(0,r) \setminus \{0\}} f \circ \Phi_N(x) \, dx = r^N \int_{B(0,1) \setminus \{0\}} f \circ \Phi_N(x) \, dx$ and therefore that

$$(IV.2.1) \qquad \int_{B(0,r) \setminus \{0\}} f \circ \Phi_N(x) \, dx = \frac{r^N}{N} \int_{\mathbf{S}^{N-1}} f(\omega) \, \lambda_{\mathbf{S}^{N-1}}(d\omega)$$

for every non-negative measurable f on $(\mathbf{S}^{N-1}, \mathcal{B}_{\mathbf{S}^{N-1}})$. In particular, using $\omega_{N-1} \equiv \lambda_{\mathbf{S}^{N-1}}(\mathbf{S}^{N-1})$ to denote the **surface area** of $\mathbf{S}^{N-1}$, we have that $\frac{\omega_{N-1}}{N}$ is the **volume** Ω_N of the unit ball $B(0,1)$ in $\mathbb{R}^N$.

Next, define $\Psi_N : (0, \infty) \times \mathbf{S}^{N-1} \longrightarrow \mathbb{R}^N \setminus \{0\}$ by $\Psi_N(r, \omega) = r\omega$. Note that Ψ_N is one-to-one and onto; the pair $(r, \omega) \equiv \Psi_N^{-1}(x) = (|x|, \Phi_N(x))$ are called the **polar coordinates** of the point $x \in \mathbb{R}^N \setminus \{0\}$. Finally, define the measure R_N on $\big((0, \infty), \mathcal{B}_{(0,\infty)}\big)$ by $R_N(\Gamma) = \int_\Gamma r^{N-1} \, dr$.

The importance of these considerations is contained in the following result.

IV.2.2 Theorem. *Referring to the preceding, one has that*

$$\lambda_{\mathbb{R}^N} = \Psi_N^* \big(R_N \times \lambda_{\mathbf{S}^{N-1}} \big) \quad \text{on} \quad \mathcal{B}_{\mathbb{R}^N \setminus \{0\}}.$$

In particular, if f is a non-negative, measurable function on $(\mathbb{R}^N, \mathcal{B}_{\mathbb{R}^N})$, then

(IV.2.3)
$$
\begin{aligned}
\int_{\mathbb{R}^N} f(x) \, dx &= \int_{(0,\infty)} r^{N-1} \left(\int_{\mathbf{S}^{N-1}} f(r\omega) \, \lambda_{\mathbf{S}^{N-1}}(d\omega) \right) dr \\
&= \int_{\mathbf{S}^{N-1}} \left(\int_{(0,\infty)} f(r\omega) r^{N-1} \, dr \right) \lambda_{\mathbf{S}^{N-1}}(d\omega).
\end{aligned}
$$

PROOF: By Exercise III.3.21 and Theorem III.4.5, all that we have to do is check that the first equation in (IV.2.3) holds for every

$$f \in C_c(\mathbb{R}^N) \equiv \{ f \in C(\mathbb{R}^N) : f \equiv 0 \text{ off of some compact set} \}.$$

To this end, let $f \in C_c(\mathbb{R}^N)$ be given and set $F(r) = \int_{B(0,r)} f(x) \, dx$ for $r > 0$. Then, by (IV.2.1), for all $r, h > 0$:

$$
\begin{aligned}
F(r+h) - F(r) &= \int_{B(0,r+h) \setminus B(0,r)} f(x) \, dx \\
&= \int_{B(0,r+h) \setminus B(0,r)} f \circ \Psi_N(r, \Phi_N(x)) \, dx \\
&\qquad + \int_{B(0,r+h) \setminus B(0,r)} \big(f(x) - f \circ \Psi_N(r, \Phi_N(x)) \big) \, dx \\
&= \frac{(r+h)^N - r^N}{N} \int_{\mathbf{S}^{N-1}} f \circ \Psi_N(r, \omega) \, \lambda_{\mathbf{S}^{N-1}}(d\omega) + o(h),
\end{aligned}
$$

where "$o(h)$" denotes a function which tends to 0 faster than h. Hence, F is continuously differentiable on $(0, \infty)$ and its derivative at $r \in (0, \infty)$ is given by $r^{N-1} \int_{\mathbf{S}^{N-1}} f \circ \Psi_N(r, \omega) \, \lambda_{\mathbf{S}^{N-1}}(d\omega)$. Since $F(r) \longrightarrow 0$ as $r \searrow 0$, the desired result now follows from the Fundamental Theorem of Calculus. ∎

IV.2.4 Exercise.

i) Show that if $\Gamma \neq \emptyset$ is an open subset of $\mathbf{S}^{N-1}$, then $\lambda_{\mathbf{S}^{N-1}}(\Gamma) > 0$.

Next, show that $\lambda_{\mathbf{S}^{N-1}}$ is **rotation invariant**. That is, show that if $\mathcal{O}$ is an $N \times N$-orthogonal matrix and $T_{\mathcal{O}}$ is the associated transformation on $\mathbb{R}^N$ (cf. the paragraph preceding Theorem II.2.2), then $(T_{\mathcal{O}})^* \lambda_{\mathbf{S}^{N-1}} = \lambda_{\mathbf{S}^{N-1}}$. Finally, use this fact to show that

$$(\text{IV.2.5 (i)}) \qquad \int_{\mathbf{S}^{N-1}} \boldsymbol{\xi} \cdot \omega \, \lambda_{\mathbf{S}^{N-1}}(d\omega) = 0$$

and

$$(\text{IV.2.5 (ii)}) \qquad \int_{\mathbf{S}^{N-1}} (\boldsymbol{\xi} \cdot \omega)(\boldsymbol{\eta} \cdot \omega) \, \lambda_{\mathbf{S}^{N-1}}(d\omega) = \Omega_N \boldsymbol{\xi} \cdot \boldsymbol{\eta}$$

for any $\boldsymbol{\xi}, \boldsymbol{\eta} \in \mathbb{R}^N$.

Hint: In proving these, let $\boldsymbol{\xi} \in \mathbb{R}^N \setminus \{0\}$ be given and consider the rotation $R_{\boldsymbol{\xi}}$ which sends $\boldsymbol{\xi}$ to $-\boldsymbol{\xi}$ but acts as the identity on the orthogonal complement of $\boldsymbol{\xi}$.

ii) Define $\Phi : [0, 2\pi] \longrightarrow \mathbf{S}^1$ by $\Phi(\theta) = \begin{bmatrix} \cos\theta \\ \sin\theta \end{bmatrix}$, and set $\mu = \Phi^* \lambda_{\mathbb{R}}\big|_{[0,2\pi]}$. Given any rotation invariant finite measure ν on $(\mathbf{S}^1, \mathcal{B}_{\mathbf{S}^1})$, show that $\nu = \frac{\nu(\mathbf{S}^1)}{2\pi} \mu$. In particular, conclude that $\lambda_{\mathbf{S}^1} = \mu$. (Cf. Exercise IV.3.15 below.)

Hint: Define

$$R_\theta = \begin{bmatrix} \cos\theta & -\sin\theta \\ \sin\theta & \cos\theta \end{bmatrix} \quad \text{for} \quad \theta \in [0, 2\pi],$$

and note that

$$\int_{\mathbf{S}^1} f \, d\nu = \frac{1}{2\pi} \int_{[0,2\pi]} \left(\int_{\mathbf{S}^1} f \circ T_{R_\theta}(\omega) \, \nu(d\omega) \right) d\theta.$$

iii) For $N \in \mathbb{Z}^+$, define $\Xi_N : [-1, 1] \times \mathbf{S}^{N-1} \longrightarrow \mathbf{S}^N$ by

$$\Xi_N(\rho, \omega) = \begin{bmatrix} (1 - \rho^2)^{1/2} \omega \\ \rho \end{bmatrix},$$

and let

$$\mu_N(\Gamma) = \int_\Gamma (1 - \rho^2)^{\frac{N}{2} - 1} \lambda_{\mathbb{R}} \times \lambda_{\mathbf{S}^{N-1}}(d\rho \times d\omega)$$

for $\Gamma \in \mathcal{B}_{[-1,1]} \times \mathcal{B}_{\mathbf{S}^{N-1}}$. Show that $\lambda_{\mathbf{S}^N} = \Xi_N^* \mu_N$.

(**Hint:** Consider

$$\int_{(0,\infty)} r^N \left(\int_{[-1,1]\times \mathbf{S}^{N-1}} f\left(r\,\Xi_N(\rho,\omega)\right) \mu_N(d\rho \times d\omega) \right) dr$$

for continuous $f : \mathbb{R}^{N+1} \longrightarrow \mathbb{R}$ with compact support.)

Finally, use this result to show that

$$\int_{\mathbf{S}^N} f(\boldsymbol{\theta}\cdot\omega)\,\lambda_{\mathbf{S}^N}(d\omega) = \boldsymbol{\omega}_{N-1} \int_{[-1,1]} f(\rho)\left(1-\rho^2\right)^{\frac{N}{2}-1} d\rho$$

for all $\boldsymbol{\theta} \in \mathbf{S}^N$ and all measurable f on $\left([-1,1],\mathcal{B}_{[-1,1]}\right)$ which are either bounded or non-negative.

IV.2.6 Exercise.

i) Justify GAUSS's trick:

$$\left(\int_{\mathbb{R}} e^{-|x|^2/2}\,dx\right)^2 = \int_{\mathbb{R}^2} e^{-|x|^2/2}\,dx = 2\pi \int_{(0,\infty)} r e^{-r^2/2}\,dr = 2\pi$$

and conclude that for any $N \in \mathbb{Z}^+$ symmetric $N \times N$-matrix A which is strictly positive definite (i.e., all the eigenvalues of A are strictly positive),

(IV.2.7) $$\int_{\mathbb{R}^N} \exp\left[-\frac{1}{2}(x, A^{-1}x)_{\mathbb{R}^N}\right] dx = (2\pi)^{N/2}\left(\det(A)\right)^{1/2}.$$

Hint: try the change of variable $\Phi(x) = T_{A^{-1/2}}x$.

ii) Define $\Gamma(\gamma) = \int_{(0,\infty)} t^{\gamma-1}e^{-t}\,dt$ for $\gamma \in (0,\infty)$. Show that, for any $\gamma \in (0,\infty)$, $\Gamma(\gamma+1) = \gamma\Gamma(\gamma)$. Also, show that $\Gamma(1/2) = \pi^{1/2}$. The function $\Gamma(\cdot)$ is called the EULER's **Gamma-function**. Notice that it provides an extension of the factorial function in the sense that $\Gamma(n+1) = n!$ for integers $n \geq 0$.

iii) Show that

$$\boldsymbol{\omega}_{N-1} = \frac{2(\pi)^{N/2}}{\Gamma(N/2)},$$

and conclude that the volume $\boldsymbol{\Omega}_N$ of the N-dimensional unit ball is given by

$$\boldsymbol{\Omega}_N = \frac{\pi^{N/2}}{\Gamma\left(\frac{N}{2}+1\right)}.$$

iv) Given $\alpha, \beta \in (0, \infty)$, show that

$$\int_{(0,\infty)} t^{-1/2} \exp\left[-\alpha^2 t - \frac{\beta^2}{t}\right] dt = \frac{\pi^{1/2} e^{-2\alpha\beta}}{\alpha}.$$

Finally, use the preceding to show that

$$\int_{(0,\infty)} t^{-3/2} \exp\left[-\alpha^2 t - \frac{\beta^2}{t}\right] dt = \frac{\pi^{1/2} e^{-2\alpha\beta}}{\beta}.$$

Hint: Define $\psi(s)$ for $s \in \mathbb{R}$ to be the unique $t \in (0, \infty)$ satisfying $s = \alpha t^{1/2} - \beta t^{-1/2}$ and use part **i)** of Exercise IV.1.6 to show that

$$\int_{(0,\infty)} t^{-1/2} \exp\left[-\alpha^2 t - \frac{\beta^2}{t}\right] dt = \frac{e^{-2\alpha\beta}}{\alpha} \int_{\mathbb{R}} e^{-s^2} ds.$$

IV.3. Jacobi's Transformation and Surface Measure.

We begin this section by deriving JACOBI's famous generalization to non-linear maps of the result in Theorem II.2.2. We will then apply JACOBI's result to show that LEBESGUE's *measure can be differentiated across a smooth surface.*

Given an open set $G \subseteq \mathbb{R}^N$ and a $\Phi \in C^1(G; \mathbb{R}^N)$ (the space of once continuously differentiable functions on G into $\mathbb{R}^N$), we define the **Jacobian matrix** $J\Phi(x) = \frac{\partial \Phi}{\partial x}(x)$ of Φ at x to be the $N \times N$-matrix whose $(i,j)^{\text{th}}$ entry is $\frac{\partial \Phi_i}{\partial x_j}(x)$. In addition, we call $\delta\Phi(x) \equiv |\det(J\Phi(x))|$ the **Jacobian** of Φ at x.

IV.3.1 Lemma. *Let G be an open set in $\mathbb{R}^N$ and Φ an element of $C^1(G; \mathbb{R}^N)$ whose JACOBIan never vanishes on G. Then Φ maps open subsets of G into open sets, and $|\Phi^{-1}(\Gamma)|_e = 0$ if $\Gamma \subseteq \Phi(G)$ and $|\Gamma|_e = 0$. In particular, Φ is measurable from $\left(G, \overline{\mathcal{B}}_{\mathbb{R}^N}[G]\right)$ onto $\left(\Phi(G), \overline{\mathcal{B}}_{\mathbb{R}^N}[\Phi(G)]\right)$.*

PROOF: By the Inverse Function Theorem, for each $x \in G$ there is an open neighborhood $U \subseteq G$ of x such that $\Phi|_U$ is invertible and its inverse has first derivatives which are bounded and continuous. Hence, G can be written as the union of a countable number of open sets on each of which Φ admits an inverse having bounded continuous first order derivatives; and so, without loss in generality, we may and will assume that Φ admits such an inverse on G itself. But, in this case: it is obvious that Φ preserves open sets; by Lemma II.2.1 applied to Φ^{-1}, we know that $|\Phi^{-1}(\Gamma)|_e = 0$ whenever $\Gamma \subseteq \Phi(G)$ and $|\Gamma|_e = 0$; and from this and Lemma II.2.1 it follows that Φ is measurable from $\left(G, \overline{\mathcal{B}}_{\mathbb{R}^N}[G]\right)$ into $\left(\Phi(G), \overline{\mathcal{B}}_{\mathbb{R}^N}[\Phi(G)]\right)$. ∎

A continuously differentiable map Φ on an open set $U \subseteq \mathbb{R}^N$ into $\mathbb{R}^N$ is called a **diffeomorphism** if it is **injective** (i.e., one-to-one) and $\delta\Phi$ never vanishes. If Φ is a diffeomorphism on the open set U and if $W = \Phi(U)$, then we say that Φ is **diffeomorphic from** U **onto** W. Also, given any set $\Gamma \subseteq \mathbb{R}^N$ and $\delta > 0$, we use

$$\Gamma^{(\delta)} \equiv \big\{ x \in \mathbb{R}^N : |y - x| < \delta \text{ for some } y \in \Gamma \big\}$$

to denote the **open δ-hull of Γ.**

IV.3.2 Theorem. (Jacobi's Transformation Formula) *Let G be an open set in $\mathbb{R}^N$ and Φ an element of $C^2(G; \mathbb{R}^N)$. If the Jacobian of Φ never vanishes, then for every measurable function f on $\big(\Phi(G), \overline{\mathcal{B}}_{\mathbb{R}^N}[\Phi(G)]\big)$, $f \circ \Phi$ is measurable on $\big(G, \overline{\mathcal{B}}_{\mathbb{R}^N}[G]\big)$ and*

$$(\text{IV.3.3}) \qquad \int_{\Phi(G)} f(y)\, dy \le \int_G f \circ \Phi(x)\, \delta\Phi(x)\, dx$$

whenever f is non-negative. Moreover, if, in addition, Φ is one-to-one on G, then (IV.3.3) can be replaced by

$$(\text{IV.3.4}) \qquad \int_{\Phi(G)} f(y)\, dy = \int_G f \circ \Phi(x)\, \delta\Phi(x)\, dx.$$

PROOF: We first note that (IV.3.4) is a consequence of (IV.3.3) when Φ is one-to-one. Indeed, if Φ is one-to-one, then the Inverse Function Theorem guarantees that $\Phi^{-1} \in C^2(\Phi(G); \mathbb{R}^N)$. In addition,

$$J\Phi^{-1}(y) = \Big(J\Phi\big(\Phi^{-1}(y)\big) \Big)^{-1} \quad \text{for} \quad y \in \Phi(G).$$

Hence we can apply (IV.3.3) to Φ^{-1} and thereby obtain

$$\int_G f \circ \Phi(x)\, \delta\Phi(x)\, dx \le \int_{\Phi(G)} f(y)\, (\delta\Phi) \circ \Phi^{-1}(y)\, \delta\Phi^{-1}(y)\, dy = \int_{\Phi(G)} f(y)\, dy;$$

which, in conjunction with (IV.3.3) yields (IV.3.4).

We next note that is suffices to prove (IV.3.3) under the assumptions that G is bounded, Φ on G has bounded second order derivatives, and $\delta\Phi$ is uniformly positive on G. In fact, if this is not already the case, then we can choose a non-decreasing sequence of bounded open sets G_n so that $\Phi|_{G_n}$ has these properties for each $n \ge 1$ and $G_n \nearrow G$. Clearly, the result for Φ on G follows from the result for $\Phi|_{G_n}$ on G_n for every $n \ge 1$. Thus, from now on, we assume that G is bounded, the second derivatives of Φ are bounded, and $\delta\Phi$ is uniformly positive on G.

Let $Q = Q(c; r) = \prod_1^N [c_i - r, c_i + r] \subseteq G$. Then, by TAYLOR's Theorem, there is an $L \in [0, \infty)$ (depending only on the bound on the second derivatives of Φ) such that

$$\Phi(Q(c; r)) \subseteq \Phi(c) + \left(T_{J\Phi(c)} Q(0; r) \right)^{(Lr^2)}.$$

(Cf. Section II.2 for the notation here.) At the same time, there is an $M < \infty$ (depending only on L, the lower bound on $\delta\Phi$, and the upper bounds on the first derivatives of Φ) such that

$$\left(T_{J\Phi(c)} Q(0; r) \right)^{(Lr^2)} \subseteq T_{J\Phi(c)} Q(0, r + Mr^2).$$

Hence, by Theorem II.2.2,

$$\left| \Phi(Q) \right| \leq \delta\Phi(c) \left| Q(0, r + Mr^2) \right| = (1 + Mr)^N \delta\Phi(c) |Q|.$$

Now define $\mu(\Gamma) = \int_\Gamma \delta\Phi(x)\, dx$ for $\Gamma \in \overline{\mathcal{B}}_{\mathbb{R}^N}[G]$, and set $\nu = \Phi^*\mu$. Given an open set $H \subseteq \Phi(G)$, use Lemma II.1.10 to choose, for each $m \in \mathbb{Z}^+$, a countable, non-overlapping, exact cover $\mathcal{C}_m$ of $\Phi^{-1}(H)$ by cubes Q with $\mathrm{diam}(Q) < 1/m$. Then, by the preceding paragraph,

$$|H| \leq \sum_{Q \in \mathcal{C}_m} |\Phi(Q)| \leq \left(1 + \frac{M}{m} \right)^N \sum_{Q \in \mathcal{C}_m} \delta\Phi(c_Q)|Q|,$$

where c_Q denotes the center of the cube Q. After letting $m \to \infty$ in the preceding, we conclude that $|H| \leq \nu(H)$ for open $H \subseteq \Phi(G)$; and so, by Exercise III.1.9, it follows that

(IV.3.5) $|\Gamma| \leq \nu(\Gamma)$ for all $\Gamma \in \overline{\mathcal{B}}_{\mathbb{R}^N}[\Phi(G)]$.

Starting from (IV.3.5), working first with simple functions, and then passing to monotone limits, we now conclude that (IV.3.4) holds for all non-negative, measurable functions f on $(\Phi(G), \overline{\mathcal{B}}_{\mathbb{R}^N}[\Phi(G)])$. ∎

As an essentially immediate consequence of Theorem IV.3.2, we have the following.

IV.3.6 Corollary. *Let G be an open set in $\mathbb{R}^N$ and $\Phi \in C^2(G; \mathbb{R}^N)$ a one-to-one map whose JACOBIan never vanishes. Set*

$$\mu_\Phi(\Gamma) = \int_\Gamma \delta\Phi(x)\, dx \quad \text{for} \quad \Gamma \in \overline{\mathcal{B}}_{\mathbb{R}^N}[G].$$

Then $\Phi^*\mu_\Phi$ coincides with the restriction $\lambda_{\Phi(G)}$ of $\lambda_{\mathbb{R}^N}$ to $\overline{\mathcal{B}}_{\mathbb{R}^N}[\Phi(G)]$. In particular,

$f \in L^1(\Phi(G), \mathcal{B}_{\mathbb{R}^N}[\Phi(G)], \lambda_{\Phi(G)})$ *if and only if* $f \circ \Phi \in L^1(G, \mathcal{B}_{\mathbb{R}^N}[G], \mu_\Phi)$,

in which case (IV.3.4) holds.

As a mnemonic device, it is useful to represent the conclusion of Corollary IV.3.6 as the statement

$$f(y)\,dy = f \circ \Phi(x)\,\delta\Phi(x)\,dx \quad \text{when} \quad y = \Phi(x).$$

We now want to apply JACOBI's formula to show how to *differentiate* LEBESGUE*'s measure across a smooth surface.* To be precise, assume that $N \geq 2$ and let G be a bounded open set in $\mathbb{R}^N$ for which there exists a smooth function $F \in C^3(\mathbb{R}^N; \mathbb{R}^N)$ with the properties that

a) $G = \{x \in \mathbb{R}^N : F(x) < 0\}$,

b) the gradient ∇F of F vanishes at no point where F itself vanishes.

(We will think of ∇F as a row vector.) Note that the boundary ∂G of G coincides with $\{x \in \mathbb{R}^N : F(x) = 0\}$.

For each $x \in \partial G$, denote by $\mathbf{T}_x \partial G$ the set of $\mathbf{v} \in \mathbb{R}^N$ for which there exists a continuously differentiable curve γ, defined on a neighborhood of $0 \in \mathbb{R}$ into ∂G, such that $\gamma(0) = x$ and $\dot{\gamma}(0) = \mathbf{v}$ ($\dot{\gamma}$ is used here to denote the derivative of γ). The set $\mathbf{T}_x(\partial G)$ is called the **tangent space to ∂G at the point x**.

IV.3.7 Lemma. *For each $x \in \partial G$, $\mathbf{T}_x \partial G$ is an $N - 1$-dimensional subspace of $\mathbb{R}^N$. Moreover, there is a unique $\mathbf{n}(x) \in \mathbf{S}^{N-1}$ which is perpendicular to $\mathbf{T}_x \partial G$ and has the property that $x + t\mathbf{n}(x) \notin \overline{G}$ for any sufficiently small $t > 0$. In fact, $\mathbf{n}(x) = \frac{\nabla F(x)}{|\nabla F(x)|}$; and there exists a $\delta > 0$ such that, for each $x \in \partial G$,*

$$x + t\mathbf{n}(x) \in \begin{cases} G & \text{if} \quad t \in (-\delta, 0) \\ \overline{G}^{\complement} & \text{if} \quad t \in (0, \delta). \end{cases}$$

Finally, for any $\delta > 0$,

$$(\partial G)^{(\delta)} = \{x + t\mathbf{n}(x) : x \in \partial G \text{ and } |t| < \delta\}.$$

PROOF: To see that $\nabla F(x) \perp \mathbf{T}_x \partial G$, let $\mathbf{v} \in \mathbf{T}_x \partial G$ be given and choose γ accordingly. Then, because $\gamma(\cdot) \in \partial G$,

$$\big(\nabla F(x), \mathbf{v}\big)_{\mathbb{R}^N} = \frac{d}{dt} F \circ \gamma(t)\big|_{t=0} = 0.$$

Conversely, if $\mathbf{v} \in \mathbb{R}^N$ satisfying $\big(\mathbf{v}, \nabla F(x)\big)_{\mathbb{R}^N} = 0$ is given, set

$$K(y) = \mathbf{v} - \frac{\big(\mathbf{v}, \nabla F(y)\big)_{\mathbb{R}^N}}{|\nabla F(y)|^2} \nabla F(y).$$

for y in an open neighborhood U of x where $|\nabla F|$ does not vanish. By the basic existence theory for ordinary differential equations, we can then find an $\epsilon > 0$ and a continuously differentiable curve $\gamma : (-\epsilon, \epsilon) \longrightarrow U$ such that $\gamma(0) = x$ and $\dot{\gamma}(t) = K(\gamma(t))$ for all $t \in (-\epsilon, \epsilon)$. Clearly $\dot{\gamma}(0) = \mathbf{v}$, and it is an easy matter to check that

$$\frac{d}{dt}\big(F \circ \gamma(t)\big) = \big(\nabla F(\gamma(t)), \dot{\gamma}(t)\big)_{\mathbb{R}^N} = 0 \quad \text{for} \quad t \in (-\epsilon, \epsilon).$$

Hence, $\mathbf{v} \in \mathbf{T}_x \partial G$.

We have now proved that $\mathbf{T}_x \partial G$ is the perpendicular complement of $\nabla F(x)$. Hence, if $\omega \in \mathbf{S}^{N-1}$ is perpendicular to $\mathbf{T}_x \partial G$, then $\omega = \pm \frac{\nabla F(x)}{|\nabla F(x)|}$. Moreover, by TAYLOR's Theorem, there is a $C \in (0, \infty)$ such that

$$\big| F(x + t\omega) - t\big(\omega, \nabla F(x)\big)_{\mathbb{R}^N} \big| \leq Ct^2, \quad (x,t) \in \partial G \times [-1, 1];$$

from which it is clear not only that $\mathbf{n}(x) = \frac{\nabla F(x)}{|\nabla F(x)|}$ but also that

$$x + t\mathbf{n}(x) \in \begin{cases} G & \text{if} \quad 0 < t < 1/C \\ \overline{G}^{\bullet} & \text{if} \quad -1/C < t < 0. \end{cases}$$

Finally, for any $\delta > 0$, it is obvious that $x + t\mathbf{n}(x) \in (\partial G)^{(\delta)}$ if $x \in \partial G$ and $|t| < \delta$. On the other hand, if $y \in (\partial G)^{(\delta)}$ and $x \in \partial G$ is chosen so that $|y - x| = |y - \partial G|$, set $\mathbf{w} = y - x$. Clearly, all that we have to do is check that $\mathbf{w} \perp \mathbf{T}_x \partial G$. To this end, let $\mathbf{v} \in \mathbf{T}_x \partial G$ and choose γ accordingly. Then $t \longmapsto |y - \gamma(t)|^2$ has a minimum at $t = 0$, and therefore

$$2(\mathbf{w}, \mathbf{v})_{\mathbb{R}^N} = 2\big(y - \gamma(0), \dot{\gamma}(0)\big)_{\mathbb{R}^N} = \frac{d}{dt}|y - \gamma(t)|^2\big|_{t=0} = 0. \quad \blacksquare$$

The vector $\mathbf{n}(x)$ is called the **outer normal** to ∂G at $x \in \partial G$. Given $\Gamma \subseteq \partial G$ and $\delta > 0$ define

$$\Gamma(\delta) = \big\{ x + t\mathbf{n}(x) : x \in \Gamma \text{ and } |t| < \delta \big\}.$$

By Lemma II.2.1, it is clear that $\Gamma(\delta) \in \overline{\mathcal{B}}_{\mathbb{R}^N}$ if $\Gamma \in \mathcal{B}_{\partial G}$. What we want to do in the remainder of this section is show that there is a measure $\lambda_{\partial G}$ on $(\partial G, \mathcal{B}_{\partial G})$ such that

$$(\text{IV.3.8}) \qquad\qquad \lambda_{\partial G}(\Gamma) = \lim_{\delta \searrow 0} \frac{1}{2\delta} \lambda_{\mathbb{R}^N}\big(\Gamma(\delta)\big)$$

for each $\Gamma \in \mathcal{B}_{\partial G}$. Note that, if it exists, $\lambda_{\partial G}$ is "the derivative of $\lambda_{\mathbb{R}^N}$ across ∂G." Also, the reader should check that the measure $\lambda_{\mathbf{S}^{N-1}}$ defined in Section IV.2 satisfies (IV.3.8) when G is the unit ball $B(0,1)$ (cf. Exercise IV.3.17 below). Thus our present notation is consistent with that used there.

The proof that the limit on the right hand side of (IV.3.8) exists and defines a measure relies on our ability to, at least locally, *flatten* ∂G. That is, we will choose local coordinates so that ∂G looks like a piece of $\mathbb{R}^{N-1}$ lying in $\mathbb{R}^N$ in such a way that G is *below* and $\overline{G}^{\complement}$ is *above*; all of which is made possible by the Inverse Function Theorem.

In the following we will be using the notation $F_{,i}$ to denote the partial derivative $\frac{\partial F}{\partial x_i}$. More generally, given a differentiable mapping Ψ on some open subset of $\mathbb{R}^m$ into $\mathbb{R}^n$, we will use the notation

$$\Psi_{,j}(u) = \frac{\partial \Psi}{\partial u_j}(u) \equiv \begin{bmatrix} \frac{\partial \Psi_1}{\partial u_j}(u) \\ \vdots \\ \frac{\partial \Psi_n}{\partial u_j}(u) \end{bmatrix} \in \mathbb{R}^m, \qquad 1 \le j \le n.$$

IV.3.9 Lemma. *Let $x^0 \in \partial G$ be given. Then there is an open set $\widehat{W}$ in $\mathbb{R}^{N-1}$ and a twice continuously differentiable injection $\Psi : \widehat{W} \longrightarrow \partial G$ such that $x^0 \in \Psi(\widehat{W})$ and $\{\Psi_{,1}(u), \ldots, \Psi_{,N-1}(u)\}$ spans $\mathbf{T}_{\Psi(u)}\partial G$ for each $u \in \widehat{W}$. Moreover, if Ψ is any continuously differentiable mapping from an open subset $\widehat{W}$ of $\mathbb{R}^{N-1}$ into ∂G and if $\Phi : \widehat{W} \times \mathbb{R} \longrightarrow \mathbb{R}^N$ is defined by*

(IV.3.10) $$\Phi(u,t) = \Psi(u) + t\mathbf{n}(\Psi(u)),$$

then $\delta\Phi(u,0) = \delta\Psi(u)$ where $\delta\Psi(u)$ is defined to be the square root of

$$\det \left[\Big(\Psi_{,i}(u), \Psi_{,j}(u)\Big)_{\mathbb{R}^N} \right]_{1 \le i,j \le N-1}$$

$$= \det \left(\begin{bmatrix} \Psi_{,1}(u)^T \\ \vdots \\ \Psi_{,N-1}(u)^T \end{bmatrix} \begin{bmatrix} \Psi_{,1}(u) & \ldots & \Psi_{,N-1}(u) \end{bmatrix} \right).$$

(When $\mathbf{v}$ is a column vector we use $\mathbf{v}^T$ to denote the corresponding row vector.) In particular, for any $u \in \widehat{W}$, there is an open neighborhood of $(u,0)$ in $\mathbb{R}^N$ on which the Φ in (IV.3.10) is a diffeomorphism if and only if

$$\{\Psi_{,1}(u), \ldots, \Psi_{,N-1}(u)\} \quad spans \quad \mathbf{T}_{\Psi(u)}\partial G.$$

PROOF: We begin by proving the second part. Thus, let Ψ be a continuously differentiable map from an open subset $\widehat{W}$ in $\mathbb{R}^{N-1}$ to ∂G and define Φ accordingly as in (IV.3.10). Then

$$J\Phi(u,0) = \begin{bmatrix} \Psi_{,1}(u) & \cdots & \Psi_{,N-1} & \mathbf{n}(\Psi(u))^T \end{bmatrix}.$$

Since the vectors $\boldsymbol{\Psi}_{,j}(u)$ are all elements of $\mathbf{T}_{\boldsymbol{\Psi}(u)}\partial G$, it is clear from the above that $\delta\Phi(u,0) > 0$ if and only if the $\boldsymbol{\Psi}_{,j}$'s span $\mathbf{T}_{\boldsymbol{\Psi}(u)}\partial G$ and that

$$\left(\delta\Phi(u,0)\right)^2 = \det\left(\begin{bmatrix} \boldsymbol{\Psi}_{,1}(u)^T \\ \vdots \\ \boldsymbol{\Psi}_{,N-1}(u)^T \\ \mathbf{n}\bigl(\boldsymbol{\Psi}(u)\bigr) \end{bmatrix} \begin{bmatrix} \boldsymbol{\Psi}_{,1}(u) & \cdots & \boldsymbol{\Psi}_{,N-1}(u) & \mathbf{n}\bigl(\boldsymbol{\Psi}(u)\bigr)^T \end{bmatrix}\right),$$

which (because $\mathbf{n}\bigl(\boldsymbol{\Psi}(U)\bigr) \perp \mathbf{T}_{\boldsymbol{\Psi}(u)}\partial G$) is clearly equal to $\bigl(\delta\boldsymbol{\Psi}(u)\bigr)^2$. With this information at hand, the rest of the second part is a simple application of the Inverse Function Theorem.

We turn now to the first assertion; and, without loss in generality, we will assume that $F_{,N}(x^0) \neq 0$. Consider the map $\Xi : \mathbb{R}^N \longrightarrow \mathbb{R}^{N-1} \times \mathbb{R}$ given by

$$\Xi(y) = \begin{bmatrix} \overline{y} - \overline{x}^0 \\ F(y) \end{bmatrix} \quad \text{where} \quad \overline{y} \equiv \begin{bmatrix} y_1 \\ \vdots \\ y_{N-1} \end{bmatrix}.$$

Clearly $\Xi(x^0) = \begin{bmatrix} \mathbf{0} \\ 0 \end{bmatrix}$ and

$$J\Xi(y) = \begin{bmatrix} \mathbf{I}_{\mathbb{R}^{N-1}} & \mathbf{0} \\ \nabla_{\overline{y}}F(y) & F_{,N}(y) \end{bmatrix},$$

where $\mathbf{I}_{\mathbb{R}^{N-1}}$ denotes the identity matrix on $\mathbb{R}^{N-1}$ and $\nabla_{\overline{y}}F$ denotes the row vector $(F_{,1},\ldots,F_{,N-1})$. Hence $\delta\Xi(x^0) > 0$, and so, by the Inverse Function Theorem, there is an open neighborhood U of x^0 in $\mathbb{R}^N$ on which Ξ is diffeomorphic. Obviously, $\Xi_N(x) = 0$ for all $x \in U \cap \partial G$, and $\Xi(x^0) = 0$. Now choose an open neighborhood $\widehat{W}$ of the origin in $\mathbb{R}^{N-1}$ and a $\delta > 0$ so that $\widehat{W} \times (-\delta,\delta) \subseteq \Xi(U)$, and let $\boldsymbol{\Psi}(u) = \Xi^{-1}(u,0)$, $u \in \widehat{W}$. Clearly $\boldsymbol{\Psi}$ is a twice continuously differentiable injection from $\widehat{W}$ into ∂G. In addition, $\boldsymbol{\Psi}_{,j}(u)$, $1 \leq j \leq N-1$, is the jth column of the non-degenerate matrix of $J\Xi^{-1}(u,0)$, and therefore the vectors $\boldsymbol{\Psi}_{,j}(u)$, $1 \leq j \leq N-1$, must form a basis in $\mathbf{T}_{\boldsymbol{\Psi}(u)}\partial G$. $\blacksquare$

Starting with Lemmas IV.3.9 and IV.3.7 and applying the usual HEINE-BOREL reasoning, we now see that there are bounded, open sets $U_1,\ldots,U_M$ in $\mathbb{R}^N$, open sets $\widehat{W}_1,\ldots,\widehat{W}_M$ in $\mathbb{R}^{N-1}$, positive numbers $\delta_1,\ldots,\delta_M$, and twice continuously differentiable maps $\boldsymbol{\Psi}^m : \widehat{W}_m \longrightarrow \partial G$, $1 \leq m \leq M$, such that

1) $\partial G \subseteq \bigcup_1^M U_m$;

2) for each $1 \le m \le M$, Ψ^m and all its derivatives of first and second order are bounded and continuous;

3) for each $1 \le m \le M$, the map Φ^m defined from Ψ^m as in (IV.3.10) is diffeomorphic from $W_m \equiv \widehat{W}_m \times (-\delta_m, \delta_m)$ onto U_m, and

$$\Phi^m(u,t) \in \begin{cases} G & \text{if } t \in (-\delta_m, 0) \\ \overline{G}^\complement & \text{if } t \in (0, \delta_m) \end{cases} \quad \text{for all } u \in \widehat{W}_m.$$

Next, define $h : \partial G \longrightarrow (0, \infty)$ by

$$h(x) = \max_{1 \le m \le M} \operatorname{dist}(x, U_m^\complement).$$

Since h is continuous and ∂G is compact,

$$\delta_0 \equiv \frac{1}{4} \min_{x \in \partial G} h(x) > 0.$$

Moreover, if

$$V_m \equiv \left\{ x \in U_m : \operatorname{dist}(x, U_m^\complement) > 3\delta_0 \right\} \quad \text{for} \quad 1 \le m \le M,$$

then $\{V_m\}_1^M$ is an open cover of ∂G and $\overline{V}_m^{(3\delta_0)} \subseteq U_m$ for each $1 \le m \le M$.

IV.3.12 Lemma. *Let δ_0 be as in the preceding. Then, for each $y \in (\partial G)^{(\delta_0)}$ there exists a unique $x \in \partial G$ and a unique $t \in (-\delta_0, \delta_0)$ such that $y = x + t\mathbf{n}(x)$. Hence, if $f : \partial G \longrightarrow \overline{\mathbb{R}}$, then there is a unique function $\tilde{f} : (\partial G)^{(\delta_0)} \longrightarrow \overline{\mathbb{R}}$ such that $\tilde{f}(x + t\mathbf{n}(x)) = f(x)$ for all $x \in \partial G$ and $t \in (-\delta_0, \delta_0)$. Furthermore, $\tilde{f}$ is measurable on $((\partial G)^{(\delta_0)}, \mathcal{B}_{\mathbb{R}^N}[(\partial G)^{(\delta_0)}])$ if f is measurable on $(\partial G, \mathcal{B}_{\partial G})$.*

PROOF: In view of Lemma IV.3.7, the first statement will be proved once we show that for $x, x' \in \partial G$ and $t, t' \in (-\delta_0, \delta_0)$, $x + t\mathbf{n}(x) \ne x' + t'\mathbf{n}(x')$ if $x \ne x'$. To this end, suppose that $x \in \partial G$, $x' \in \partial G \setminus \{x\}$, and $t, t' \in (-\delta_0, \delta_0)$ are given; and set $y = x + t\mathbf{n}(x)$ and $y' = x' + t'\mathbf{n}(x')$. If $|x - x'| > 2\delta_0$, then

$$2\delta_0 < |x - x'| \le |x - y| + |y - y'| + |y' - x'| < 2\delta_0 + |y - y'|,$$

and so $y \ne y'$. On the other hand, if $|x - x'| \le 2\delta_0$, then there is an $1 \le m \le M$ such that $x = \Phi^m(u, 0)$, $x' = \Phi^m(u', 0)$, $y = \Phi^m(u, t)$, and $y' = \Phi^m(u', t')$, where $(u, 0)$, $(u', 0)$, (u, t), and (u', t') are all elements of W_m and $u \ne u'$. Since Φ^m is one-to-one on W_m, it follows that $y \ne y'$.

As for the second assertion, it is obvious that $\tilde{f}$ is well-defined for any function f on ∂G. To see that if f is measurable on $(\partial G, \mathcal{B}_{\partial G})$ then $\tilde{f}$ is measurable on $(\partial G^{(\delta_0)}, \mathcal{B}_{\mathbb{R}^N}[(\partial G)^{(\delta_0)}])$, it suffices to show that, for each $1 \leq m \leq M$, the restriction of $\tilde{f}$ to $(\partial G)^{(\delta_0)} \cap V_m$ is measurable. But, for $x \in (\partial G)^{(\delta_0)} \cap V_m$,

$$\tilde{f}(x) = [f \circ \Psi^m]\Big((\Phi^m)_1^{-1}(x), \ldots, (\Phi^m)_{N-1}^{-1}(x)\Big). \quad \blacksquare$$

We can now prove that $\lambda_{\partial G}$ exists.

IV.3.13 Theorem. *There is a unique measure $\lambda_{\partial G}$ on $(\partial G, \mathcal{B}_{\partial G})$ such that (IV.3.8) holds for every $\Gamma \in \mathcal{B}_{\partial G}$. In fact, $\lambda_{\partial G}$ is a finite measure, and if f is a bounded measurable function on $(\partial G, \mathcal{B}_{\partial G})$ then*

$$\int_{\partial G} f \, d\lambda_{\partial G} = \lim_{\delta \searrow 0} \frac{1}{2\delta} \int_{(\partial G)^{(\delta)}} \tilde{f}(x) \, dx,$$

where $\tilde{f}$ is the function described in Lemma IV.3.12.

Finally, let Ψ be a twice continuously differentiable injection from an open set $\widehat{W}$ of $\mathbb{R}^{N-1}$ into ∂G, and assume that $\{\Psi_{,1}(u), \ldots, \Psi_{,N-1}(u)\}$ is a basis for $\mathbf{T}_{\Psi(u)} \partial G$ at each $u \in \widehat{W}$. Then $\Psi(\widehat{W})$ is open in ∂G and, for every bounded or non-negative, measurable function f on $(\partial G, \mathcal{B}_{\partial G})$,

$$(IV.3.14) \qquad \int_{\Psi(\widehat{W})} f(x) \, \lambda_{\partial G}(dx) = \int_{\widehat{W}} f \circ \Psi(u) \, \delta\Psi(u) \, du,$$

where $\delta\Psi(u)$ is defined as in the second part of Lemma IV.3.9.

PROOF: Let Ψ on $\widehat{W}$ be as in the last part of the statement, and define Φ from Ψ as in (IV.3.10). By Lemma IV.3.12, Φ is injective on $\widehat{W} \times (-\delta_0, \delta_0)$; and, by the last part of Lemma IV.3.9, it is therefore diffeomorphic on an open W in $\mathbb{R}^N$ with $\widehat{W} = \{u \in \mathbb{R}^{N-1} : (u, 0) \in W\}$. Hence, $\Psi(\widehat{W}) = \Phi(W) \cap \partial G$ is open in ∂G, and, by Theorem IV.3.2, for any bounded, measurable function f on $(\partial G, \mathcal{B}_{\partial G})$, we have that

$$\int_{\Phi(W) \cap (\partial G)^{(\delta)}} \tilde{f}(x) \, dx = \int_W f \circ \Psi(u) \, \chi_{(-\delta,\delta)}(t) \, \delta\Phi(u,t) \, du \times dt, \quad \delta \in (0,\infty).$$

In particular, by FUBINI's Theorem combined with LEBESGUE's Dominated Convergence Theorem, we see, from the second part of Lemma IV.3.9, that

$$\lim_{\delta \searrow 0} \frac{1}{2\delta} \int_{\Phi(W) \cap \partial G^{(\delta)}} \tilde{f}(x) \, dx = \int_{\widehat{W}} f \circ \Psi(u) \, \delta\Psi(u) \, du.$$

Now refer to the discussion preceding Lemma IV.3.12, and define $A_1 = U_1 \cap \partial G$ and

$$A_m = (U_m \cap \partial G) \setminus \left(\bigcup_{\ell=1}^{m-1} A_\ell \right) \quad \text{for} \quad 2 \le \ell \le M;$$

and set $B_m = (\Psi^m)^{-1}(A_m)$. By the preceding paragraph, one then has that, for any bounded, measurable f on $(\partial G, \mathcal{B}_{\partial G})$:

$$\lim_{\delta \searrow 0} \frac{1}{2\delta} \int_{(\partial G)^{(\delta)}} \tilde{f}(x) \, dx = \sum_{m=1}^{M} \int_{B_m} f \circ \Psi^m(u) \, \delta\Psi(u) \, du;$$

and clearly this not only proves that the limit in (IV.3.8) exists but also that the resulting quantity $\lambda_{\partial G}$ is a finite measure on $(\partial G, \mathcal{B}_{\partial G})$. In addition, after combining the preceding with the result obtained in the first part of this proof, we see that (IV.3.14) holds first for all bounded and then for all non-negative functions f which are measurable on $(\partial G, \mathcal{B}_{\partial G})$. ∎

The measure $\lambda_{\partial G}$ produced in Theorem IV.3.13 is called the **surface measure on** ∂G.

IV.3.15 Exercise.

In the final assertion of part ii) in Exercise IV.2.4 and again in i) of Exercise IV.2.6, we tacitly accepted the equality of π, the volume Ω_2 of the unit ball $B(0,1)$ in $\mathbb{R}^2$, with π, the half-period of the sin and cos functions. We are now in a position to justify this identification. To this end, define $\Phi : G \equiv (0,1) \times (0, 2\pi) \longrightarrow \mathbb{R}^2$ (the π here is the half-period of sin and cos) by $\Phi(r, \theta) = (r \cos\theta, r \sin\theta)$. Note that $\Phi(G) = B(0,1) \setminus \{x \in B(0,1) : x_1 = 0\}$ and therefore that $|\Phi(G)| = \Omega_2$. Now use JACOBI's Transformation Formula to compute $|\Phi(G)|$.

IV.3.16 Exercise.

Let G be an open subset in $\mathbb{R}^N$ of the sort discussed in Lemma IV.3.7. Show that, for each $x \in \partial G$, the tangent space $\mathbf{T}_x(\partial G)$ coincides with the set of $\mathbf{v} \in \mathbb{R}^N$ such that

$$\varlimsup_{\xi \to \infty} \frac{\text{dist}(x + \xi\mathbf{v}, \partial G)}{\xi^2} < \infty.$$

Hint: Given $\mathbf{v} \in \mathbf{T}_x(\partial G)$, show that the corresponding curve γ may be chosen to be twice continuously differentiable and consider $\xi \longmapsto |x + \xi\mathbf{v} - \gamma(\xi)|$.

IV.3.17 Exercise.

Given $r > 0$, define $\Phi_r : \mathbf{S}^{N-1} \longrightarrow \partial B(0,r)$ by $\Phi_r(\omega) = r\omega$. Show that the surface measure $\lambda_{\partial B(0,r)}$ on $\partial B(0,r)$ coincides with $r^{N-1} \cdot \Phi^* \lambda_{\mathbf{S}^{N-1}}$, where $\lambda_{\mathbf{S}^{N-1}}$ is the measure described in Section IV.2. In particular, this means that the measure in Section IV.2 is the surface measure on $\partial B(0,1)$.

IV.3.18 Exercise.

Show that if $\Gamma \neq \emptyset$ is an open subset of ∂G, then $\lambda_{\partial G}(\Gamma) > 0$.

IV.3.19 Exercise.

i) For $(\alpha, \beta) \in (0, \infty)^2$, define

$$B(\alpha, \beta) = \int_{(0,1)} u^{\alpha-1}(1-u)^{\beta-1} \, du.$$

Show that

$$B(\alpha, \beta) = \frac{\Gamma(\alpha)\Gamma(\beta)}{\Gamma(\alpha+\beta)}$$

where $\Gamma(\cdot)$ is the Gamma-function described in **ii)** of Exercise IV.2.6. (**Hint:** Think of $\Gamma(\alpha)\,\Gamma(\beta)$ as an integral in (s,t) over $(0,\infty)^2$ and consider the map

$$(u, v) \in (0, \infty) \times (0, 1) \longmapsto \big(uv, u(1-v)\big) \in (0, \infty)^2.)$$

The function B is called the **Beta-function**. Clearly it provides an extension of the binomial coefficients in the sense that

$$\frac{m+n+1}{B(m+1, n+1)} = \binom{m+n}{m}$$

for all non-negative integers m and n.

ii) For $\lambda > N/2$ show that

$$\int_{\mathbb{R}^N} \frac{1}{(1+|x|^2)^\lambda} \, dx = \frac{\pi^{N/2}\Gamma(\lambda - N/2)}{\Gamma(\lambda)}.$$

(**Hint:** use polar coordinates and then try the change of variable $\Phi(r) = \frac{r^2}{1+r^2}$.) In particular, conclude that

$$\int_{\mathbb{R}^N} \frac{1}{(1+|x|^2)^{(N+1)/2}} = \frac{\omega_N}{2},$$

where ω_N is the surface area of the sphere $\mathbf{S}^N$ in $\mathbb{R}^{N+1}$ (cf. part **iii)** of Exercise IV.2.6.)

iii) For $\lambda \in (0, \infty)$, show that

$$\int_{(-1,1)} (1 - \xi^2)^{\lambda-1} \, d\xi = \frac{\pi^{1/2} \Gamma(\lambda)}{\Gamma\left(\lambda + \frac{1}{2}\right)};$$

and conclude that, for any $N \in \mathbb{Z}^+$,

$$\int_{(-1,1)} (1 - \xi^2)^{\frac{N}{2}-1} \, d\xi = \frac{\omega_N}{\omega_{N-1}}.$$

Finally, check that this last result is consistent with part **iii)** of Exercise IV.2.4.

IV.4. The Divergence Theorem.

Again let $N \geq 2$ and $G = \{x \in \mathbb{R}^N : F(x) < 0\}$ be as in our discussion of surface measure (cf. the paragraph preceding Lemma IV.3.7). Our main goal in this section is to prove that if $H : \mathbb{R}^N \longrightarrow \mathbb{R}^N$ is a smooth function, then

(IV.4.1) $$\int_G \text{div} H(x) \, dx = \int_{\partial G} (H(x), \mathbf{n}(x))_{\mathbb{R}^N} \, \lambda_{\partial G}(dx),$$

where $\text{div} H = \sum_{I=1}^N H_{i,i}$ is the **divergence** of H and we continue to use the notation introduced in the last section for partial derivatives (thus, $H_{i,i} = \frac{\partial H_i}{\partial x_i}$).

The key to (IV.4.1) is contained in the following lemma.

IV.4.2 Lemma. *Let* $\omega \in \mathbf{S}^{N-1}$. *If* $f \in C_c(G)$ *(i.e.,* f *is continuous and vanishes off of a compact subset of* G*), then*

$$\frac{d}{d\xi} \int_G f(x + \xi\omega) \, dx \bigg|_{\xi=0} = 0.$$

On the other hand, if U_m *is one of the open sets described in the paragraph preceding Lemma IV.3.12 and if* $f \in C_c(U_m)$, *then*

$$\frac{d}{d\xi} \int_G f(x + \xi\omega) \, dx \bigg|_{\xi=0} = \int_{\partial G} f(x) \, (\mathbf{n}(x), \omega)_{\mathbb{R}^N} \, \lambda_{\partial G}(dx).$$

PROOF: If $f \in C_c(V_0)$, then for all sufficiently small $|\xi|$:

$$\int_G f(x + \xi\omega) \, dx = \int_{\mathbb{R}^N} f(x + \xi\omega) \, dx = \int_{\mathbb{R}^N} f(x) \, dx = \int_G f(x) \, dx,$$

and therefore $\frac{d}{d\xi} \int_G f(x + \xi\omega) \, dx \bigg|_{\xi=0} = 0.$

Now suppose that $1 \leq m \leq M$ and that $f \in C_c(U_m)$. Again referring to the paragraph preceding Lemma IV.3.12, one has that, for sufficiently small non-zero $|\xi|$'s:

$$\int_G f(x + \xi\omega)\,dx - \int_G f(x)\,dx = \int_{U_m} \Big(\chi_G(x - \xi\omega) - \chi_G(x)\Big) f(x)\,dx$$

$$= \int_{W_m} \Big[\chi_G\big(\Phi^m(u,t) - \xi\omega\big) - \chi_G\big(\Phi^m(u,t)\big)\Big] f\big(\Phi^m(u,t)\big)\,\delta\Phi^m(u,t)\,du \times dt$$

$$= I(\xi) + J(\xi),$$

where $I(\xi)$ and $J(\xi)$ are defined, respectively, to be

$$\int_{W_m} \Big[\chi_G\big(\Phi^m(u,t) - \xi\omega\big)$$
$$- \chi_G\Big(\Phi^m(u,t) - \xi\omega(u)\mathbf{n}\big(\Psi^m(u)\big)\Big)\Big] f\big(\Phi^m(u,t)\big)\delta\Phi^m(u,t)\,du \times dt$$

and

$$\int_{W_m} \Big[\chi_G\Big(\Phi^m(u,t) - \xi\omega(u)\mathbf{n}\big(\Psi^m(u)\big)\Big)$$
$$- \chi_G\big(\Phi^m(u,t)\big)\Big] f\big(\Phi^m(u,t)\big)\delta\Phi^m(u,t)\,du \times dt,$$

with

$$\omega(u) \equiv \Big(\mathbf{n}\big(\Psi^m(u)\big), \omega\Big)_{\mathbb{R}^N}.$$

Notice that another expression for $J(\xi)$ is

$$\int_{W_m} \Big[\chi_{(-\infty,0)}\big(t - \xi\omega(u)\big) - \chi_{(-\infty,0)}(t)\Big] f\big(\Phi^m(u,t)\big)\,\delta\Phi^m(u,t)\,du \times dt$$

$$= \int_{\widehat{W}_m^+(\xi)} \left(\int_{(0,\xi\omega(u))} f\big(\Phi^m(u,t)\big)\,\delta\Phi^m(u,t)\,dt\right) du$$

$$- \int_{\widehat{W}_m^-(\xi)} \left(\int_{(-\xi\omega(u),0)} f\big(\Phi^m(u,t)\big)\,\delta\Phi^m(u,t)\,dt\right) du,$$

where $\widehat{W}_m^{\pm}(\xi) \equiv \big\{u \in \widehat{W}_m : \pm\xi\omega(u) > 0\big\}$; and therefore

$$\lim_{\xi \to 0} \frac{J(\xi)}{\xi} = \int_{\widehat{W}_m} f\big(\Psi^m(u)\big)\,\omega(u)\,\delta\Psi^m(u)\,du = \int_{\partial G} f(x)\big(\mathbf{n}(x),\omega\big)_{\mathbb{R}^N}\,\lambda_{\partial G}(dx).$$

In view of the preceding, all that remains is to show that $\lim_{\xi \to 0} \frac{I(\xi)}{\xi} = 0$. To this end, first choose a compact subset $\hat{K}$ of $\widehat{W}_m$ and a $\delta \in (0, \delta_m)$ so that f vanishes off of $\Phi^m(\hat{K} \times (-\delta, \delta))$, and set $\alpha = \frac{1}{2}\text{dist}(\Phi^m(\hat{K}) \times (-\delta, \delta), U_m^{\complement})$. Next, for $u \in \hat{K}$, define $S_+(u, \xi)$ and $S_-(u, \xi)$ to be the sets

$$\left\{ t : |t| < \delta, \ \Phi^m(u, t) - \xi\omega \in G \text{ and } \Phi^m(u, t) - \omega(u)\mathbf{n}(\Psi^m(u)) \notin G \right\}$$

and

$$\left\{ t : |t| < \delta, \ \Phi^m(u, t) - \xi\omega \notin G \text{ and } \Phi^m(u, t) - \omega(u)\mathbf{n}(\Psi^m(u)) \in G \right\},$$

respectively. It is then clear that $I(\xi)$ is dominated by a constant times

$$\int_{\hat{K}} \lambda_{\mathbb{R}}(S_+(u, \xi) \cup S_-(u, \xi)) \, du;$$

and therefore, by LEBESGUE's Dominated Convergence Theorem, we will be done once we show that

$$\lim_{\xi \to 0} \frac{\lambda_{\mathbb{R}}(S_+(u, \xi) \cup S_-(u, \xi))}{\xi} = 0 \quad \text{for each} \quad u \in \hat{K}.$$

Thus, let $u \in \hat{K}$ be given, and set

$$x = \Psi^m(u), \quad y(t, \xi) = \Phi^m(u, t) - \xi\omega, \quad \text{and} \quad z(t, \xi) = \Phi^m(u, t) - \xi\omega(u)\mathbf{n}(x).$$

Clearly, so long as $|\xi| < \alpha$ and $|t| < \delta$, both $y(t, \xi)$ and $z(t, \xi)$ are elements of U_m. Next, set $\mathbf{v} = \omega - \omega(u)\mathbf{n}(x)$ and $s(t, \xi) = t - \xi\omega(u)$. Then,

$$y(t, \xi) = x - \xi\mathbf{v} + s(t, \xi)\mathbf{n}(x) \quad \text{and} \quad z(t, \xi) = x + s(t, \xi)\mathbf{n}(x).$$

Moreover, when $|\xi| < \alpha$, there is a unique $x(\xi) \in U_m \cap \partial G$ and $\tau(\xi) \in \mathbb{R}$ such that $x - \xi\mathbf{v} = x(\xi) + \tau(\xi)\mathbf{n}(x(\xi))$. In fact, because $\mathbf{v} \in \mathbf{T}_x(\partial G)$, (cf. Exercise IV.3.16) there is a $C_1 \in (0, \infty)$ such that

$$|\tau(\xi)| = \text{dist}(x - \xi\mathbf{v}, U_m \cap \partial G) \le C_1\xi^2.$$

Because, at the same time,

$$|\mathbf{n}(x) - \mathbf{n}(x(\xi))| \le C_2|x - x(\xi)| \le C_2(|\xi| + C_1\xi^2)$$

for some $C_2 \in (0, \infty)$, we now see that there is a $C_3 \in (0, \infty)$ such that

$$\left| y(t, \xi) - \Big(x(\xi) + s(t, \xi)\mathbf{n}(x(\xi)) \Big) \right| \leq C_3 \Big(|s(t, \xi)| \, |\xi| + \xi^2 \Big)$$

for $|\xi| < \alpha$ and $|t| < \delta$. In particular, we can find a $0 < \rho \leq \alpha$ such that $x(\xi) + s(t, \xi)\mathbf{n}(x) \in U_m$ and therefore

$$|s(t, \xi)| = \mathrm{dist}\big(x(\xi) + s(t, \xi)\mathbf{n}(x), \partial G\big) \quad \text{for} \quad |\xi| \leq \rho \text{ and } |t| < \delta.$$

Now, suppose that $|\xi| \leq \rho$ and $t \in S_+(u, \xi)$. Then, $z(t, \xi) \in U_m \cap G^\complement$, and therefore $s(t, \xi) \geq 0$. On the other hand, $y(t, \xi) \in U_m \cap G$, and therefore, by the preceding,

$$\begin{aligned}
s(t, \xi) &= \mathrm{dist}\big(x(\xi) + s(t, \xi)\mathbf{n}(x), \partial G\big) \\
&\leq \left| y(t, \xi) - \Big(x(\xi) + s(t, \xi)\mathbf{n}(x(\xi))\Big) \right| \leq C_3\Big(s(t, \xi)|\xi| + \xi^2\Big);
\end{aligned}$$

from which it follows that $s(t, \xi) \leq C\xi^2$ for some $C \in (0, \infty)$. In other words, we have now shown that

$$S_+(u, \xi) \subseteq \big(\xi\omega(u), \xi\omega(u) + C\xi^2\big) \quad \text{for} \quad |\xi| \leq \rho.$$

Since the same argument leads to

$$S_-(u, \xi) \subseteq \big(\xi\omega(u) - C\xi^2, \xi\omega(u)\big) \quad \text{for} \quad |\xi| \leq \rho,$$

we conclude that $\lambda_{\mathbb{R}}\big(S_+(u) \cup S_-(u)\big) \leq 2C\xi^2$, which is more than enough to get the desired result. ∎

IV.4.3 Theorem. (THE DIVERGENCE THEOREM) *If f has continuous first order derivatives in a neighborhood of $\overline{G}$, then, for any $\omega \in \mathbb{R}^N$,*

$$(\mathrm{IV.4.4}) \qquad \int_G \big(\nabla f(x), \omega\big)_{\mathbb{R}^N} \, dx = \int_{\partial G} f(x) \, \big(\mathbf{n}(x), \omega\big)_{\mathbb{R}^N} \lambda_{\partial G}(dx).$$

Thus, if H is an $\mathbb{R}^N$-valued function which has continuous first derivatives in a neighborhood of $\overline{G}$, then (IV.4.1) holds.

PROOF: It suffices to prove the first assertion; and, in doing so, we may and will assume that $\omega \in \mathbf{S}^{N-1}$.

Note that

$$\int_G (\nabla f(x), \omega)_{\mathbb{R}^N} \, dx = \frac{d}{d\xi} \int_G f(x + \xi\omega) \, dx \bigg|_{\xi=0}.$$

Thus, all that we have to do is check that

$$\frac{d}{d\xi} \int_G f(x + \xi\omega) \, dx \bigg|_{\xi=0} = \int_{\partial G} f(x) \, (\mathbf{n}(x), \omega)_{\mathbb{R}^N} \, \lambda_{\partial G}(dx).$$

Since, by Lemma IV.4.2, the preceding equation holds if f is an element of $C_c(G) \cup \bigcup_{m=1}^{M} C_c(U_m)$, we will be done once we show that there exist $\eta_0, \ldots, \eta_M$ such that $\eta_0 \in C_c(G)$, $\eta_m \in C_c(U_m)$ for $1 \leq m \leq M$, and $\sum_{m=0}^{M} \eta_m \equiv 1$ in a neighborhood of $\overline{G}$. Indeed, we can then write $f = \sum_{m=0}^{M} \eta_m f$ in a neighborhood of $\overline{G}$, and can therefore obtain the desired result from the fact that it holds for each of the functions $\eta_m f$. To construct the η_m's, let $\{V_m\}_1^M$ be the open cover of ∂G discussed just before Lemma IV.3.12, and choose an open V_0 so that $\overline{V}_0 \subseteq G$ and $\overline{G} \subseteq \bigcup_{m=0}^{M} V_m$. Finally, choose an open D so that

$$\overline{G} \subseteq D \subseteq \overline{D} \subseteq \bigcup_{m=0}^{M} V_m,$$

and define

$$\psi_m(x) = \text{dist}(x, V_m^{\complement}) \quad \text{and} \quad s(x) = \text{dist}(x, \overline{D}) + \sum_{m=0}^{M} \psi_m(x).$$

It is then clear that $\psi_0 \in C_c(G)$, $\psi_m \in C_c(U_m)$ for $1 \leq m \leq M$, and that $s \geq \epsilon$ for some $\epsilon > 0$. Hence $\eta_m \equiv \frac{\psi_m}{s} \in C_c(U_m)$ and, in addition, $\sum_{m=0}^{M} \eta_m \equiv 1$ on D. ∎

IV.4.5 Remark.

The Divergence Theorem may be thought of as an integration by parts formula in which the derivatives on the integrand have been transferred to LEBESGUE's measure, resulting in an integral with respect to surface measure. (In fact, as the reader can easily check, the one-dimensional version of The Divergence Theorem is the Fundamental Theorem of Calculus.) The fact that it is the LEBESGUE's measure which is being differentiated in the passage from the left to the right side of (IV.4.4) is highlighted by the observation that the equation

$$\frac{d}{d\xi} \int_G f(x + \xi\omega) \, dx \bigg|_{\xi=0} = \int_{\partial G} f(x) (\mathbf{n}(x), \omega)_{\mathbb{R}^N} \, \lambda_{\partial G}(dx)$$

does *not* require f to be differentiable.

Before dropping this topic, we will give some examples of the way in which The Divergence Theorem is used in the analysis of partial differential equations.

Let $\Delta = \sum_{i=1}^{N} \frac{\partial^2}{\partial x_i^2}$ denote the standard **Laplacian**. The following variant on The Divergence Theorem provides one of the keys to the analysis of equations in which Δ appears.

IV.4.6 Theorem. (GREEN'S IDENTITY) *Let u and v be smooth functions in a neighborhood of $\overline{G}$. Then*

(IV.4.7)
$$\int_G u \, \Delta v \, dx - \int_G v \, \Delta u \, dx$$
$$= \int_{\partial G} u \, \frac{\partial v}{\partial \mathbf{n}} \, \lambda_{\partial G}(dx) - \int_{\partial G} v \, \frac{\partial u}{\partial \mathbf{n}} \, \lambda_{\partial G}(dx),$$

where

$$\frac{\partial f}{\partial \mathbf{n}}(x) \equiv \frac{d}{d\xi} f(x + \xi \mathbf{n}(x))\Big|_{\xi=0} = (\nabla f(x), \mathbf{n}(x))_{\mathbb{R}^N}$$

denotes differentiation in the direction $\mathbf{n}$.

PROOF: Simply note that $u \, \Delta v - v \, \Delta u = \operatorname{div}(u \, \nabla v - v \, \nabla u)$, and apply The Divergence Theorem. ∎

In order to get information out of GREEN's Identity, one must make judicious choices of v for a given u. For example, one often wants to take v to be the **fundamental solution** g given by $g(0) = \infty$ and

(IV.4.8)
$$g(x) = \begin{cases} -\log|x| & \text{if } N = 2 \\ |x|^{2-N} & \text{if } N \geq 3 \end{cases} \quad \text{for } x \in \mathbb{R}^N \setminus \{0\}.$$

Note that g and $|\nabla g|$ are integrable on every compact subset of $\mathbb{R}^N$. In addition, the following facts about g are easy to verify:

(IV.4.9)
$$\Delta g(x) = 0 \quad \text{and} \quad |x|^N \, \nabla g(x) = \begin{cases} -x & \text{if } N = 2 \\ (2-N)x & \text{if } N \geq 3 \end{cases}$$

on $\mathbb{R}^N \setminus \{0\}$.

Our first application allows us to solve the **Poisson equation** $\Delta u = -f$.

IV.4.10 Theorem. *Set* $c_N = 2\pi$ *or* $(N-2)\omega_{N-1}$ *(cf. Exercise IV.2.6) depending on whether* $N = 2$ *or* $N \geq 3$. *Given* $f \in C_c^2(\mathbb{R}^N)$, *define* u_f *on* $\mathbb{R}^N$ *by*

$$(\text{IV.4.11}) \qquad u_f(x) = \frac{1}{c_N} \int_{\mathbb{R}^N} g(x-y) f(y) \, dy.$$

Then $u_f \in C^2(\mathbb{R}^N)$ *and* $\Delta u_f = -f$.

PROOF: First observe that another expression for $u_f(\cdot)$ is $\int_{\mathbb{R}^N} g(y) f(\cdot - y) \, dy$, and it is clear from this latter expression not only that $u_f \in C^2(\mathbb{R}^N)$ but also that $\Delta u_f(x) = \int_{\mathbb{R}^N} g(y) \Delta f(x-y) \, dy$. Thus, all that we need to do is check that $\int_{\mathbb{R}^N} g(y) \Delta f(x-y) \, dy = -c_N f(x)$.

Fix x and choose $R > 1$ so that $f \equiv 0$ off of $B(x, R-1)$. For $0 < r < R$, set $G_r = B(0,R) \setminus \overline{B(0,r)}$. Then

$$\int_{\mathbb{R}^N} g(y) \Delta f(x-y) \, dy = \lim_{r \searrow 0} \int_{G_r} g(y) \Delta f(x-y) \, dy;$$

and, by GREEN's Identity and (IV.4.9), for each $0 < r < R$:

$$\int_{G_r} g(y) \Delta f(x-y) \, dy$$

$$= \int_{\partial G_r} g(y) \frac{\partial f}{\partial \mathbf{n}}(x-y) \, \lambda_{\partial G_r}(dy) - \int_{\partial G_r} f(x-y) \frac{\partial g}{\partial \mathbf{n}}(y) \, \lambda_{\partial G_r}(dy)$$

$$= - \int_{\partial B(0,r)} g(y) \frac{\partial f}{\partial \rho}(x-y) \, \lambda_{\partial B(0,r)}(dy)$$

$$\qquad + \int_{\partial B(0,r)} f(x-y) \frac{\partial g}{\partial \rho}(y) \, \lambda_{\partial B(0,r)}(dy)$$

$$= -r^{N-1} \int_{\mathbf{S}^{N-1}} g(r\omega) \frac{\partial f}{\partial \rho}(x+r\omega) \lambda_{\mathbf{S}^{N-1}}(d\omega)$$

$$\qquad + r^{N-1} \int_{\mathbf{S}^{N-1}} f(x+r\omega) \frac{\partial g}{\partial \rho}(r\omega) \lambda_{\mathbf{S}^{N-1}}(d\omega),$$

where $\frac{\partial}{\partial \rho}$ denotes differentiation in the outward radial direction and we have used Exercise IV.3.17, and the fact that, for G_r, $\frac{\partial}{\partial \mathbf{n}} = -\frac{\partial}{\partial \rho}$ on $\partial B(0,r)$. But $r^{N-1} g(r\omega) \longrightarrow 0$ uniformly as $r \searrow 0$ and

$$r^{N-1} \frac{\partial g}{\partial \rho}(r\omega) = \begin{cases} -1 & \text{if } N = 2 \\ -(N-2) & \text{if } N \geq 3. \end{cases}$$

After combining this with the preceding, we now see that

$$\lim_{r \searrow 0} \int_{G_r} g(y) \Delta f(x - y)\, dy$$

$$= -\frac{c_N}{\omega_{N-1}} \lim_{r \searrow 0} \int_{S^{N-1}} f(x + r\omega)\, \lambda_{S^{N-1}}(d\omega) = -c_N f(x). \quad\blacksquare$$

Our second application of GREEN's Identity will be to harmonic functions. A function $u \in C^2(G)$ is said to be **harmonic in** G if $\Delta u = 0$. Notice that if $N = 1$ and u is harmonic on (a, b) and continuous on $[a, b]$, then $u(x) = \frac{(b-x)}{(b-a)} u(a) + \frac{(x-a)}{(b-a)} u(b)$ for $x \in [a, b]$. In particular, $u\left(\frac{a+b}{2}\right)$ is precisely the mean of the values that u takes on $\partial(a, b)$. We will now use GREEN's Identity to derive the analogous result about harmonic functions in higher dimensions.

IV.4.12 Theorem. (THE MEAN VALUE PROPERTY) *Suppose that u is an harmonic element of $C^2(G)$. Then, for each $x \in G$ and $R > 0$ satisfying $\overline{B(x, R)} \subseteq G$,*

$$(IV.4.13) \qquad u(x) = \frac{1}{\omega_{N-1}} \int_{S^{N-1}} u(x + R\omega)\, \lambda_{S^{N-1}}(d\omega).$$

PROOF: Without loss in generality, we will assume that $x = 0$.

Set $g_R(x) = g(x) - g(R\omega^0)$ where $\omega^0 = (1, 0, \ldots, 0) \in \mathbb{R}^N$. Then, by GREEN's Identity applied to u and g_R in the region G_r used in the proof of the preceding,

$$0 = \int_{G_r} \left(g_R(y)\, \Delta u(y) - u(y)\, \Delta g_R(y) \right) dy$$

$$= -R^{N-1} \int_{S^{N-1}} u(R\omega) \frac{\partial g_R}{\partial \rho}(R\omega) \lambda_{S^{N-1}}(d\omega)$$

$$- r^{N-1} \int_{S^{N-1}} g_R(r\omega) \frac{\partial u}{\partial \rho}(r\omega) \lambda_{S^{N-1}}(d\omega)$$

$$+ r^{N-1} \int_{S^{N-1}} u(r\omega) \frac{\partial g_R}{\partial \rho}(r\omega) \lambda_{S^{N-1}}(d\omega),$$

where we have used the same notation as in the preceding proof. Note that the first term on the right equals (cf. Theorem IV.4.10)

$$\frac{c_N}{\omega_{N-1}} \int_{S^{N-1}} u(R\omega)\, \lambda_{S^{N-1}}(d\omega),$$

the second term tends to 0 as $r \searrow 0$, while the third term tends to $-c_N u(0)$. ∎

IV.4.14 Exercise.

Let u be a twice continuously differentiable function in a neighborhood of the closed ball $\overline{B(x,r)}$ and assume that $\Delta u \leq 0$ in $B(x,r)$. Generalize the Mean Value Property by showing that

$$(IV.4.15) \qquad u(x) \geq \frac{1}{\omega_{N-1}} \int_{\mathbb{S}^{N-1}} u(x + r\omega) \, \lambda_{\mathbb{S}^{N-1}}(d\omega)$$

Next, show that (IV.4.13) and (IV.4.15) yield, respectively,

$$(IV.4.16\,(i)) \qquad u(x) = \frac{1}{\Omega_N r^N} \int_{B(x,r)} u(x + y) \, dy$$

and

$$(IV.4.16\,(ii)) \qquad u(x) \geq \frac{1}{\Omega_N r^N} \int_{B(x,r)} u(x + y) \, dy.$$

Using (IV.4.16 (ii)), argue that if G is a connected open set in $\mathbb{R}^N$ and $u \in C^2(G)$ satisfies $\Delta u \leq 0$, then u achieves its minimum value at an $x \in G$ if and only if u is constant on G. This fact is known as the **strong minimum principle.**

IV.4.17 Exercise.

Let $F \in C^3(\mathbb{R}^2; \mathbb{R})$ satisfy $|F(x,y)| + |\nabla F(x,y)| > 0$, $(x,y) \in \mathbb{R}^2$, and assume that $G \equiv \{(x,y) \in \mathbb{R}^2 : F(x,y) < 0\}$ is bounded. In addition, assume that $\gamma \in C^2([0,1); \mathbb{R}^2)$ has the properties that:

$$t \in [0,1) \longmapsto \gamma(t) \in \partial G \quad \text{is an injective surjection,}$$
$$\gamma(0) = \lim_{t \nearrow 1} \gamma(t), \quad \dot{\gamma}(0) = \lim_{t \nearrow 1} \dot{\gamma}(t),$$
$$\ddot{\gamma}(0) = \lim_{t \nearrow 1} \ddot{\gamma}(t), \quad \text{and} \quad |\dot{\gamma}(t)| > 0 \quad \text{for} \quad t \in [0,1).$$

i) Show that

$$\int_{\partial G} \phi(\zeta) \, \lambda_{\partial G}(d\zeta) = \int_{[0,1]} \phi \circ \gamma(t) \, |\dot{\gamma}(t)| \, dt$$

for all bounded measurable ϕ on ∂G.

ii) Suppose that $h \in C^2(G^{(\delta)}; \mathbb{R})$ for some $\delta > 0$, and define $u = \frac{\partial h}{\partial x}$ and $v = -\frac{\partial h}{\partial y}$. Next, define $f = u + iv$ and $\Gamma(t) = \gamma_1(t) + i\gamma_2(t)$, where $i = \sqrt{-1}$. Show that

$$(IV.4.18) \qquad \int_{[0,1]} f(\gamma(t))\, \dot{\Gamma}(t)\, dt = i \int_G [\Delta h](\zeta)\, d\zeta.$$

A particularly important case of (IV.4.18) is the one when $\Delta h = 0$; in which case f is a complex analytic function and (IV.4.18) leads to the famous **Cauchy integral formula**.

IV.4.19 Exercise.

Suppose that $F : \mathbb{R}^N \longrightarrow \mathbb{R}$ is a thrice continuously differentiable function with the properties that $G(t) \equiv \{x \in \mathbb{R}^N : F(x) < t\}$ is bounded for each $t < 1$ and that ∇F vanishes nowhere on $H \equiv \{x \in \mathbb{R}^N : 0 < F(x) < 1\}$.

For each $t \in (0,1)$, let μ_t denote the surface measure $\lambda_{\partial G(t)}$ on $\partial G(t)$. Show that, for every non-negative measurable function f on H,

$$t \in (0,1) \longmapsto \int_{\partial G(t)} f(x)\, \mu_t(dx) \in [0, \infty]$$

is measurable and that

$$\int_H f(x)\, dx = \int_{(0,1)} \left(\int_{\partial G(t)} f(x)\, \mu_t(dx) \right)\, dt.$$

Chapter V. Some Basic Inequalities

V.1. Jensen, Minkowski, and Hölder.

There are a few general inequalities which play a central role in measure theory and its applications. The ones dealt with in this section are all consequences of convexity considerations.

A subset $C \subseteq \mathbb{R}^N$ is said to be **convex** if $(1-t)x + ty \in C$ whenever $x, y \in C$ and $t \in [0,1]$. Given a convex set $C \subseteq \mathbb{R}^N$, we say that $g : C \longrightarrow \mathbb{R}$ is a **concave function** on C if

$$g\big((1-t)x + ty\big) \geq (1-t)g(x) + tg(y) \quad \text{for all} \quad x, y \in C \text{ and } t \in [0,1].$$

Note that g is concave on C if and only if $\big\{(x,t) \in C \times \mathbb{R} : t \leq g(x)\big\}$ is a convex subset of $\mathbb{R}^{N+1}$.

The essence of the relationship between these notions and measure theory is contained in the following.

V.1.1 Theorem. (JENSEN'S INEQUALITY) *Let C be a closed, convex subset of $\mathbb{R}^N$ and suppose that g is a continuous, concave, non-negative function on C. Let $(E, \mathcal{B}, \mu)$ be a probability space and $F : E \longrightarrow C$ a measurable function on $(E, \mathcal{B})$ with the property that $|F| \in L^1(\mu)$. Then*

$$\int_E F \, d\mu \equiv \begin{bmatrix} \int_E F_1 \, d\mu \\ \vdots \\ \int_E F_N \, d\mu \end{bmatrix} \in C$$

and

$$\int_E g \circ F \, d\mu \leq g\left(\int_E F \, d\mu\right).$$

PROOF: First assume that F is simple. Then $F = \sum_{k=0}^n y_k \chi_{\Gamma_k}$ for some $n \in \mathbb{Z}^+$, $y_0, \ldots, y_n \in C$, and mutually disjoint sets $\Gamma_0, \ldots, \Gamma_n \in \mathcal{B}$ which cover E. Hence, since $\sum_0^n \mu(\Gamma_k) = 1$ and C is convex, $\int_E F \, d\mu = \sum_0^n y_k \mu(\Gamma_k) \in C$ and, because g is concave,

$$g\left(\int_E F \, d\mu\right) = g\left(\sum_0^n y_k \mu(\Gamma_k)\right) \geq \sum_0^n g(y_k) \mu(\Gamma_k) = \int_E g \circ F \, d\mu.$$

103

Next let F be general. Choose and fix some element y_0 of C, and let $\{y_k\}_1^\infty$ be a dense sequence in C. Given $m \in \mathbb{Z}^+$, choose $R_m > 0$ so that

$$\int_{\{|F| \geq R_m\}} (|y_0| + |F|)\, d\mu \leq \frac{1}{m}$$

and $n_m \in \mathbb{Z}^+$ so that $C \cap \overline{B(0, R_m)} \subseteq \bigcup_{k=1}^{n_m} B(y_k, 1/m)$. Define $\Gamma_{m,0} = \{\xi \in E : |F(\xi)| \geq R_m\}$, and use induction to define

$$\Gamma_{m,\ell} = \left\{ \xi \in E \setminus \bigcup_{k=0}^{\ell-1} \Gamma_{k,m} : F(\xi) \in B(y_\ell, 1/m) \right\}$$

for $1 \leq \ell \leq n_m$. Finally, set $F_m = \sum_{k=0}^{n_m} y_k \chi_{\Gamma_{m,k}}$. Then, by the preceding, $\int_E F_m\, d\mu \in C$ and

$$g\left(\int_E F_m\, d\mu \right) \geq \int_E g \circ F_m\, d\mu$$

for each $m \in \mathbb{Z}^+$. Moreover, it is easy to see that $\big\| |F_m - F| \big\|_{L^1(\mu)} \longrightarrow 0$ as $m \to \infty$. Thus, because C is closed, we now see that $\int_E F\, d\mu \in C$. At the same time, because g is continuous, $g \circ F_m \longrightarrow g \circ F$ in μ-measure as $m \to \infty$. Hence, by Fatou's Lemma

$$\int_E g \circ F\, d\mu \leq \varliminf_{m\to\infty} \int_E g \circ F_m\, d\mu \leq \varliminf_{m\to\infty} g\left(\int_E F_m\, d\mu \right) = g\left(\int_E F\, d\mu \right). \quad \blacksquare$$

We now need to develop a criterion for recognizing when a function is concave. Such a criterion is contained in the next theorem.

V.1.2 Lemma. *Suppose that $C^\circ \subseteq \mathbb{R}^N$ is an open convex set, and let C be its closure. Then C is convex. Moreover, if g is continuous on C and $g \in C^2(C^\circ)$, then g is concave on C if and only if its* **Hessian matrix**

$$H_g(x) \equiv \left[\frac{\partial^2 g}{\partial x_i \partial x_j}(x) \right]_{1 \leq i,j \leq N}$$

is non-positive definite (i.e., all of the eigenvalues of $H_g(x)$ are non-positive) for each $x \in C^\circ$.

PROOF: The convexity of C is obvious.

In order to prove that g is concave on C if $H_g(x)$ is non-positive definite at every $x \in C^\circ$, we will use the following simple result about functions on the interval $[0,1]$. Namely, suppose that u is continuous on $[0,1]$ and that u has two continuous derivatives on $(0,1)$. Then $u(t) \geq 0$ for every $t \in [0,1]$ if $u(0) = u(1) = 0$ and $u''(t) \leq 0$ for every $t \in (0,1)$. To see this, let $\epsilon > 0$ be given and consider the function $u_\epsilon \equiv u - \epsilon t(t-1)$. Clearly it is enough for us to show that $u_\epsilon \geq 0$ on $[0,1]$ for every $\epsilon > 0$. Note that $u_\epsilon(0) = u_\epsilon(1) = 0$ and $u_\epsilon''(t) < 0$ for every $t \in (0,1)$. In particular, if $u_\epsilon(t) < 0$ for some $t \in [0,1]$, then there is an $s \in (0,1)$ at which u_ϵ achieves its absolute minimum. But this is impossible, since then we would have that $u_\epsilon''(s) \geq 0$. (The astute reader will undoubtedly see that this result could have been derived as a consequence of the strong minimum principle in Exercise IV.4.14 for $N = 1$.)

Now assume that $H_g(x)$ is non-positive definite for every $x \in C^\circ$. Given $x, y \in C^\circ$, define $u(t) = g((1-t)x + ty) - (1-t)g(x) - tg(y)$ for $t \in [0,1]$. Then $u(0) = u(1) = 0$ and

$$u''(t) = \Big(y - x, H_g((1-t)x + ty)(y-x)\Big)_{\mathbb{R}^N} \leq 0$$

for every $t \in (0,1)$. Hence, by the preceding paragraph, $u \geq 0$ on $[0,1]$; and so $g((1-t)x + ty) \geq (1-t)g(x) + tg(y)$ for all $t \in [0,1]$. In other words, g is concave on C°. But, by continuity, this means that g is also concave on C.

To complete the proof, suppose that $H_g(x)$ has a positive eigenvalue for some $x \in C^\circ$. We can then find an $\omega \in S^{N-1}$ and an $\epsilon > 0$ such that $\big(\omega, H_g(x)\omega\big)_{\mathbb{R}^N} > 0$ and $x + t\omega \in C^\circ$ for all $t \in (-\epsilon, \epsilon)$. Set $u(t) = g(x + t\omega)$ for $t \in (-\epsilon, \epsilon)$. Then $u''(0) = \big(\omega, H_g(x)\omega\big)_{\mathbb{R}^N} > 0$. On the other hand,

$$u''(0) = \lim_{t \to 0} \frac{u(t) + u(-t) - 2u(0)}{t^2};$$

and if g were concave then

$$2u(0) = 2u\left(\frac{1}{2}t + \frac{1}{2}(-t)\right) = 2g\left(\frac{1}{2}(x + t\omega) + \frac{1}{2}(x - t\omega)\right)$$

$$\geq g(x + t\omega) + g(x - t\omega) = u(t) + u(-t);$$

and therefore we would have the contradiction that $u''(0) \leq 0$. ∎

V.1.3 Lemma. Let $A = \begin{bmatrix} a & b \\ b & c \end{bmatrix}$ be a real symmetric matrix. Then A is non-positive if and only if both $a + c \leq 0$ and $ac \geq b^2$. In particular, for each

$\alpha \in (0,1)$, the functions $(x,y) \in [0,\infty)^2 \longmapsto x^\alpha y^{1-\alpha}$ and $(x,y) \in [0,\infty)^2 \longmapsto (x^\alpha + y^\alpha)^{1/\alpha}$ are continuous and concave.

PROOF: In view of Lemma V.1.2, it suffices for us to check the first assertion. To this end, let $T = a + c$ be the trace and $D = ac - b^2$ the determinant of A. Also, let λ and μ denote the eigenvalues of A. Then, $T = \lambda + \mu$ and $D = \lambda\mu$.

If A is non-positive and therefore $\lambda \vee \mu \le 0$, then it is obvious that $T \le 0$ and that $D \ge 0$. If $D > 0$, then either both λ and μ are positive or they are both negative. Hence if, in addition, $T \le 0$, they must both be negative. Finally, if $D = 0$ and $T \le 0$, then either $\lambda = 0$ and $\mu = T \le 0$ or $\mu = 0$ and $\lambda = T \le 0$. ∎

V.1.4 Theorem. (MINKOWSKI'S INEQUALITY) *Let f_1 and f_2 be non-negative, measurable functions on the measure space $(E, \mathcal{B}, \mu)$. Then, for every $p \in [1,\infty)$,*

$$\left(\int_E (f_1 + f_2)^p \, d\mu \right)^{\frac{1}{p}} \le \left(\int_E f_1^p \, d\mu \right)^{\frac{1}{p}} + \left(\int_E f_2^p \, \mu \right)^{\frac{1}{p}}.$$

PROOF: The case when $p = 1$ is trivial. Also, without loss in generality, we assume that f_1^p and f_2^p are μ-integrable and that f_1 and f_2 are $[0,\infty)$-valued.

Let $p \in (1,\infty)$ be given. If we assume that $\mu(E) = 1$ and we take $\alpha = \frac{1}{p}$, then by Lemma V.1.3 and JENSEN's inequality,

$$\int_E (f_1 + f_2)^p \, d\mu = \int_E \left[(f_1^p)^\alpha + (f_2^p)^\alpha \right]^{1/\alpha} d\mu$$

$$\le \left[\left(\int_E f_1^p \, d\mu \right)^\alpha + \left(\int_E f_2^p \, d\mu \right)^\alpha \right]^{1/\alpha}$$

$$= \left[\left(\int_E f_1^p \, d\mu \right)^{1/p} + \left(\int_E f_2^p \, d\mu \right)^{1/p} \right]^p.$$

More generally, if $\mu(E) = 0$ there is nothing to do, and if $0 < \mu(E) < \infty$ we can replace μ by $\mu/\mu(E)$ and apply the preceding. Hence, all that remains is the case when $\mu(E) = \infty$. But if $\mu(E) = \infty$, take $E_n = \{f_1 \vee f_2 \ge 1/n\}$, note that $\mu(E_n) \le n^p \int f_1^p \, d\mu + n^p \int f_2^p \, d\mu < \infty$, apply the preceding to f_1, f_2, and μ all restricted to E_n, and let $n \to \infty$. ∎

V.1.5 Theorem. (HÖLDER'S INEQUALITY) *Given $p \in (1, \infty)$ define the HÖL-*
DER *conjugate p' of p by the equation $\frac{1}{p} + \frac{1}{p'} = 1$. Then, for every pair of
non-negative, measurable functions f_1 and f_2 on the measure space $(E, \mathcal{B}, \mu)$,*

$$\int_E f_1 \cdot f_2 \, d\mu \leq \left(\int_E f_1^p \, d\mu \right)^{\frac{1}{p}} \left(\int_E f_2^{p'} \, d\mu \right)^{\frac{1}{p'}}$$

for every $p \in (1, \infty)$.

PROOF: First note that if either factor on the right hand side of the above
inequality is 0, then $f_1 \cdot f_2 = 0$ (a.e., μ), and so the left hand side is also 0.
Thus we will assume that both factors on the right are strictly positive; in
which case, we may and will assume in addition that both f_1^p and $f_2^{p'}$ are
μ-integrable and that f_1 and f_2 are both $[0, \infty)$-valued. Also, just as in the
proof of MINKOWSKI's inequality, we can reduce everything to the case when
$\mu(E) = 1$. But then we can use apply JENSEN's Inequality and Lemma V.1.3
to see that

$$\int_E f_1 \cdot f_2 \, d\mu = \int_E (f_1^p)^{\alpha} \cdot (f_2^{p'})^{1-\alpha} \, d\mu \leq \left(\int_E f_1^p \, d\mu \right)^{\alpha} \left(\int_E f_2^{p'} \, d\mu \right)^{1-\alpha}$$

$$= \left(\int_E f_1^p \, d\mu \right)^{1/p} \left(\int_E f_2^{p'} \, d\mu \right)^{1/p'},$$

where $\alpha = 1/p$. ∎

V.1.6 Exercise.

i) Show that $\log$ is a continuous and concave on every interval $[\epsilon, \infty)$ with
$\epsilon > 0$. Use this together with JENSEN's inequality to show that for any $n \in \mathbb{Z}^+$,
$\mu_1, \ldots, \mu_n \in (0, 1)$ satisfying $\sum_{m=1}^n \mu_m = 1$, and $a_1, \ldots, a_n \in (0, \infty)$,

$$\prod_{m=1}^n a_m^{\mu_m} \leq \sum_{m=1}^n \mu_m a_m.$$

In particular, when $\mu_m = 1/n$ for every $1 \leq m \leq n$ this yields $\left(a_1 \cdots a_n \right)^{1/n} \leq \frac{1}{n} \sum_{m=1}^n a_m$, which is the statement that *the arithmetic mean dominates the
geometric mean.*

ii) Let $n \in \mathbb{Z}^+$, and suppose that $f_1, \ldots, f_n$ are non-negative measurable functions on the measure space $(E, \mathcal{B}, \mu)$. Given $p_1, \ldots, p_n \in (1, \infty)$ satisfying

$\sum_{m=1}^{n} 1/p_m = 1$, show that

$$\int_E f_1 \cdots f_n \, d\mu \le \prod_{m=1}^{n} \left(\int_E f_m^{p_m} \, d\mu \right)^{1/p_m}.$$

V.1.7 Exercise.

When $p = 2$, MINKOWSKI's and HÖLDER's inequalities are intimately related and are both very simple to prove. Indeed, let f_1 and f_2 be bounded, non-negative, measurable functions on the finite measure space $(E, \mathcal{B}, \mu)$. Given any $\alpha \ne 0$, observe that

$$0 \le \int_E \left(\alpha f_1 - \frac{1}{\alpha} f_2 \right)^2 \, d\mu = \alpha^2 \int_E f_1^2 \, d\mu - 2 \int_E f_1 \cdot f_2 \, d\mu + \frac{1}{\alpha^2} \int_E f_2^2 \, d\mu,$$

from which it follows that

$$2 \int_E f_1 \cdot f_2 \, d\mu \le t \int_E f_1^2 \, d\mu + \frac{1}{t} \int_E f_2^2 \, d\mu$$

for every $t > 0$. If either integral on the right vanishes, show from the preceding that $\int_E f_1 \cdot f_2 \, d\mu \le 0$. On the other hand, if neither integral vanishes, choose $t > 0$ so that the preceding yields

$$(V.1.8) \qquad \int_E f_1 \cdot f_2 \, d\mu \le \left(\int_E f_1^2 \, d\mu \right)^{1/2} \left(\int_E f_2^2 \, d\mu \right)^{1/2}.$$

Hence, in any case, (V.1.8) holds. Finally, argue that one can remove the restriction that f_1 and f_2 be bounded and then remove the condition that $\mu(E) < \infty$.

Clearly (V.1.8) is the special case of HÖLDER's inequality when $p = 2$. Because it is a particularly significant case, it is often referred to by a different name and is called **Schwarz's inequality**. Assuming that both f_1^2 and f_2^2 are μ-integrable, show that the inequality in SCHWARZ's inequality is an equality if and only if there exist $\alpha, \beta \in \mathbb{R}$ such that $\alpha f_1 + \beta f_2 = 0$ (a.e., μ).

Finally, use SCHWARZ's inequality to obtain MINKOWSKI's inequality for the case when $p = 2$. Notice the similarity between the development here and that of the classical *triangle inequality* for the EUCLIDean metric on $\mathbb{R}^N$.

V.2. The Lebesgue Spaces.

In Section III.2 we introduced $\| \cdot \|_{L^1(\mu)}$ and the space $L^1(\mu)$. We are now ready to introduce analogous quantities for other values of p.

Given a measure space $(E, \mathcal{B}, \mu)$ and a $p \in [1, \infty)$, define

$$\|f\|_{L^p(\mu)} = \left(\int_E |f|^p \, d\mu \right)^{1/p}$$

for measurable functions f on $(E, \mathcal{B})$. Also, if f is a measurable function on $(E, \mathcal{B})$ define

$$\|f\|_{L^\infty(\mu)} = \inf\Big\{ M \in [0, \infty] : |f| \leq M \ (\text{a.e.}, \mu) \Big\}.$$

Obviously, as p varies $\|f\|_{L^p(\mu)}$ provides different estimates on the size of f as it is "seen" by the measure μ.

Although information about f can be gleaned from a study of $\|f\|_{L^p(\mu)}$ as p changes (for example, *spikes* in f will be emphasized by taking p to be large), all these quantities share the same flaw as $\|f\|_{L^1(\mu)}$: they cannot detect properties of f which occur on sets having μ-measure 0. Thus, before we can hope to use any of them to get a metric on measurable functions, we must invoke the same subterfuge which we introduced at the end of Section III.2 in connection with the space $L^1(\mu)$. Namely, for $p \in [1, \infty]$, we denote by $L^p(\mu) = L^p(E, \mathcal{B}, \mu)$ the collection of equivalence classes $[f]^{\overset{\mu}{\sim}}$ (cf. the paragraph following Lemma III.2.12) of $\mathbb{R}$-valued measurable functions f satisfying $\|f\|_{L^p(\mu)} < \infty$. Once again, we will abuse notation by using f to denote its own equivalence class $[f]^{\overset{\mu}{\sim}}$.

Note that, by (III.2.13) and MINKOWSKI's inequality,

$$(\text{V.2.1}) \qquad \|\alpha f_1 + \beta f_2\|_{L^p(\mu)} \leq |\alpha| \, \|f_1\|_{L^p(\mu)} + |\beta| \, \|f_2\|_{L^p(\mu)}$$

for all $p \in [1, \infty)$, $f_1, f_2 \in L^p(\mu)$, and $\alpha, \beta \in \mathbb{R}$. Moreover, it is a simple matter to check that (V.2.1) continues to hold when $p = \infty$. Thus, each of the spaces $L^p(\mu)$ is a vector space. In addition, because of our convention and MARKOV's inequality (Theorem III.2.8), $\|f\|_{L^p(\mu)} = 0$ if and only if $f = 0$ as an element of $L^p(\mu)$. Finally, (V.2.1) allows us to check that $\|f_2 - f_1\|_{L^p(\mu)}$ satisfies the triangle inequality and, together with the preceding, this shows that $\|f_2 - f_1\|_{L^p(\mu)}$ determines a metric on $L^p(\mu)$. Thus, when $\{f_n\}_1^\infty \cup \{f\} \subseteq L^p(\mu)$, we often say $f_n \longrightarrow f$ in $L^p(\mu)$ when we mean that $\|f_n - f\|_{L^p(\mu)} \longrightarrow 0$.

The following theorem simply summarizes obvious applications of the results in Sections III.2 and III.3 to the present context. The reader should check that he sees how each of the assertions here follows from relevant result there.

V.2.2 Theorem. *Let $(E, \mathcal{B}, \mu)$ be a measure space. Then, for any measurable functions f and g on $(E, \mathcal{B})$ and $p \in [1, \infty]$,*

$$\left| \|g\|_{L^p(\mu)} - \|f\|_{L^p(\mu)} \right| \leq \|g - f\|_{L^p(\mu)}.$$

Next suppose that $\{f_n\}_1^\infty \subseteq L^p(\mu)$ for some $p \in [1, \infty]$ and that f is an $\mathbb{R}$-valued measurable function on $(E, \mathcal{B})$.

i) *If $p \in [1, \infty)$ and $f_n \longrightarrow f$ in $L^p(\mu)$, then $f_n \longrightarrow f$ in μ-measure. If $f_n \longrightarrow f$ in $L^\infty(\mu)$, then $f_n \longrightarrow f$ uniformly off of a set of μ-measure 0.*

ii) *If $p \in [1, \infty]$ and $f_n \longrightarrow f$ in μ-measure or (a.e., μ), then $\|f\|_{L^p(\mu)} \leq \underline{\lim}_{n \to \infty} \|f_n\|_{L^p(\mu)}$. Moreover, if $p \in [1, \infty)$ and, in addition, there is a $g \in L^p(\mu)$ such that $|f_n| \leq g$ (a.e., μ) for each $n \in \mathbb{Z}^+$, then $f_n \longrightarrow f$ in $L^p(\mu)$.*

iii) *If $p \in [1, \infty]$ and $\lim_{m \to \infty} \sup_{n \geq m} \|f_n - f_m\|_{L^p(\mu)} = 0$, then there is an $f \in L^p(\mu)$ such that $f_n \longrightarrow f$ in $L^p(\mu)$. In other words, the space $L^p(\mu)$ is complete with respect to the metric determined by $\| \cdot \|_{L^p(\mu)}$.*

Finally, we have the following variants of Theorem III.3.13 and Corollary III.3.14.

iv) *Assume that $\mu(E) < \infty$ and that $p, q \in [1, \infty)$. Referring to Theorem III.3.13, define S as in that theorem. Then for each $f \in L^p(\mu) \cap L^q(\mu)$ there is a sequence $\{\phi_n\}_1^\infty \subseteq S$ such that $\phi_n \longrightarrow f$ both in $L^p(\mu)$ and in $L^q(\mu)$. In particular, if $\mathcal{B}$ is generated by a countable collection $\mathcal{C}$, then each of the spaces $L^p(\mu)$, $p \in [1, \infty)$, is separable.*

v) *Let (E, ρ) be a metric space and suppose that μ is a measure on $(E, \mathcal{B}_E)$ for which there exists a non-decreasing sequence of open sets $E_n \nearrow E$ satisfying $\mu(E_n) < \infty$ for each $n \geq 1$. Then, for each pair $p, q \in [1, \infty)$ and $f \in L^p(\mu) \cap L^q(\mu)$, there is a sequence $\{\phi_n\}_1^\infty$ of bounded ρ-uniformly continuous functions such that $\phi_n \equiv 0$ off of E_n and $\phi_n \longrightarrow f$ both in $L^p(\mu)$ and in $L^q(\mu)$.*

The version of LIEB's variation on FATOU's Lemma for L^p-spaces with $p \neq 1$ is not so easy as the assertions in Theorem V.2.2. To prove it we will need the following lemma.

V.2.3 Lemma. *Let $p \in (1, \infty)$, and suppose that $\{f_n\}_1^\infty \subseteq L^p(\mu)$ satisfies $\sup_{n \geq 1} \|f_n\|_{L^p(\mu)} < \infty$ and that $f_n \longrightarrow 0$ either in μ-measure or (a.e., μ). Then, for every $g \in L^p(\mu)$,*

$$\lim_{n \to \infty} \int |f_n|^{p-1} \cdot |g| \, d\mu = 0 = \lim_{n \to \infty} \int |f_n| \cdot |g|^{p-1} \, d\mu.$$

PROOF: Without loss in generality, we assume that all of the f_n's as well as g are non-negative. Given $\delta > 0$, we have that

$$\int f_n^{p-1} \cdot g \, d\mu = \int_{\{f_n \leq \delta g\}} f_n^{p-1} \cdot g \, d\mu + \int_{\{f_n > \delta g\}} f_n^{p-1} \cdot g \, d\mu$$

$$\leq \delta^{p-1} \|g\|_{L^p(\mu)}^p + \int_{\{f_n \geq \delta^2\}} f_n^{p-1} \cdot g \, d\mu + \int_{\{g \leq \delta\}} f_n^{p-1} \cdot g \, d\mu.$$

Applying HÖLDER's inequality to each of the last two terms, we now see that

$$\int f_n^{p-1} \cdot g \, d\mu \leq \delta^{p-1} \|g\|_{L^p(\mu)}^p$$

$$+ \|f_n\|_{L^p(\mu)}^{p-1} \left[\left(\int_{\{f_n \geq \delta^2\}} g^p \, d\mu \right)^{1/p} + \left(\int_{\{g \leq \delta\}} g^p \, d\mu \right)^{1/p} \right].$$

Since, by LEBESGUE's Dominated Convergence Theorem, the first term in the final brackets tend to 0 as $n \to 0$, we conclude that

$$\varlimsup_{n \to \infty} \int f_n^{p-1} \cdot g \, d\mu \leq \delta^{p-1} \|g\|_{L^p(\mu)}^p + \sup_{n \geq 1} \|f_n\|_{L^p(\mu)}^{p-1} \|\chi_{\{g \leq \delta\}} \cdot g\|_{L^p(\mu)}$$

for every $\delta > 0$. Thus, after another application of LEBESGUE's Dominated Convergence Theorem, we get the first result upon letting $\delta \searrow 0$.

To treat the other case, apply the preceding with f_n^{p-1} and g^{p-1} replacing f_n and g, respectively, and with p' in place of p. ∎

V.2.4 Theorem. (LIEB) *Let $(E, \mathcal{B}, \mu)$ be a measure space, $p \in [1, \infty)$, and $\{f_n\}_1^\infty \cup \{f\} \subseteq L^p(\mu)$. If $\sup_{n \geq 1} \|f_n\|_{L^p(\mu)} < \infty$ and $f_n \longrightarrow f$ in μ-measure or (a.e., μ), then*

$$(V.2.5) \qquad \lim_{n \to \infty} \int \left| |f_n|^p - |f|^p - |f_n - f|^p \right| d\mu = 0.$$

In particular, $\|f_n - f\|_{L^p(\mu)} \longrightarrow 0$ if $\|f_n\|_{L^p(\mu)} \longrightarrow \|f\|_{L^p(\mu)}$.

PROOF: The case when $p = 1$ is the content of Theorems III.3.5 and III.3.12, and so we will assume that $p \in (1, \infty)$. Given such a p, we first check that there is a $K_p < \infty$ such that

$$(V.2.6) \quad \left| |b|^p - |a|^p - |b - a|^p \right| \leq K_p \left(|b - a|^{p-1} |a| + |a|^{p-1} |b - a| \right), \quad a, b \in \mathbb{R}.$$

Since (V.2.6) clearly holds for all $a, b \in \mathbb{R}$ if it does for all $a \in \mathbb{R} \setminus \{0\}$ and $b \in \mathbb{R}$, we can divide both sides of (V.2.6) by $|a|^p$ and thereby show that (V.2.6) is equivalent to

$$\big| \, |c|^p - 1 - |c - 1|^p \, \big| \leq K_p \left(|c - 1|^{p-1} + |c - 1| \right), \quad c \in \mathbb{R};$$

and the existence of a $K_p < \infty$ for which this inequality holds can be easily verified with elementary considerations of what happens when c is near 1 and when $|c|$ is near infinity.

Applying (V.2.6) with $a = f_n(x)$ and $b = f(x)$, we see that

$$\big| \, |f_n|^p - |f|^p - |f_n - f|^p \, \big| \leq K_p \left(|f_n - f|^{p-1} \cdot |f| + |f_n - f| \cdot |f|^{p-1} \right)$$

pointwise. Thus, by Lemma V.2.3 with f_n and g there replaced by $f_n - f$ and f, respectively, our result follows. ∎

We now turn to the application of HÖLDER's inequality to the L^p-spaces. In order to do so, we first complete the definition of the HÖLDER conjugate p' which, thus far, has only been defined (cf. Theorem V.1.5) for $p \in (1, \infty)$. Thus, we define $p' = \infty$ or 1 according to whether $p = 1$ or ∞. Notice that this is completely consistent with the equation $\frac{1}{p} + \frac{1}{p'} = 1$ used before.

V.2.7 Theorem. *Let $(E, \mathcal{B}, \mu)$ be a measure space.*

i) *If f and g are measurable functions on $(E, \mathcal{B})$, then for every $p \in [1, \infty]$*

(V.2.8) $$\|f \cdot g\|_{L^1(\mu)} \leq \|f\|_{L^p(\mu)} \|g\|_{L^{p'}(\mu)}.$$

In particular, if $f \in L^p(\mu)$ and $g \in L^{p'}(\mu)$, then $f \cdot g \in L^1(\mu)$.

ii) *If $p \in [1, \infty)$ and $f \in L^p(\mu)$, then*

(V.2.9) $$\|f\|_{L^p(\mu)} = \sup \left\{ \int f \cdot g \, d\mu : g \in L^{p'}(\mu) \text{ and } \|g\|_{L^{p'}(\mu)} \leq 1 \right\}.$$

In fact, if $\|f\|_{L^p(\mu)} > 0$, then the supremum in (V.2.9) is achieved by the function

$$g = \frac{|f|^{p-1} \cdot \text{sgn} \circ f}{\|f\|_{L^p(\mu)}^{p-1}}.$$

iii) *More generally, for any f which is measurable on $(E, \mathcal{B})$*

(V.2.10) $$\|f\|_{L^p(\mu)} = \sup \left\{ \|f \cdot g\|_{L^1(\mu)} : g \in L^{p'}(\mu) \text{ and } \|g\|_{L^{p'}(\mu)} \leq 1 \right\}$$

if $p = 1$ or if $p \in (1, \infty)$ and either $\mu(|f| \geq \delta) < \infty$ for every $\delta > 0$ or μ is σ-finite.

PROOF: Part i) is an immediate consequence of HÖLDER's inequality when $p \in (1, \infty)$. At the same time, when $p \in \{1, \infty\}$, the conclusion is clear without any further comment. Given i), ii) is easy.

When $p = 1$, **iii**) is obvious; and, in view of **ii**), the proof of **iii**) for $p \in (1, \infty)$ reduces to showing that, under either one of the stated conditions, $\|f\|_{L^p(\mu)} = \infty$ implies that the right hand side of (V.2.10) is infinite. To this end, first suppose that $\mu(|f| \geq \delta) < \infty$ for every $\delta > 0$. Then, for each $n \geq 1$, the function

$$\psi_n \equiv |f|^{p-1} \cdot \left(\chi_{[1/n,n]} \circ |f| \right) \in L^{p'}(\mu).$$

Moreover, if $\|f\|_{L^p(\mu)} = \infty$, then, by the Monotone Convergence Theorem, $\|\psi_n\|_{L^{p'}(\mu)} \longrightarrow \infty$. Thus, since $\|f \cdot \psi_n\|_{L^1(\mu)} = \|\psi_n\|_{L^{p'}(\mu)}^{p'}$, we see that

$$\|f \cdot g_n\|_{L^1(\mu)} \longrightarrow \infty \quad \text{if} \quad \|f\|_{L^p(\mu)} = \infty \quad \text{and} \quad g_n \equiv \frac{\psi_n}{1 + \|\psi_n\|_{L^{p'}(\mu)}}.$$

Finally, suppose that μ is σ-finite and that $\mu(|f| \geq \delta) = \infty$ for some $\delta > 0$. Choose $\{E_n\}_1^\infty \subseteq \mathcal{B}$ so that $E_n \nearrow E$ and $\mu(E_n) < \infty$ for every $n \geq 1$. Then it is easy to see that $\lim_{n \to \infty} \|f \cdot g_n\|_{L^1(\mu)} = \infty$ when

$$g_n \equiv \frac{\chi_{\Gamma_n}}{\left(1 + \mu(\Gamma_n)\right)^{1-1/p}} \quad \text{with} \quad \Gamma_n = E_n \cap \{|f| \geq \delta\}.$$

Since $\|g_n\|_{L^{p'}(\mu)} \leq 1$, this completes the proof. ∎

For reasons which will become clearer in the next section, it is sometimes useful to consider the following slight variation on the basic L^p-spaces. Namely, let $(E_1, \mathcal{B}_1, \mu_1)$ and $(E_2, \mathcal{B}_2, \mu_2)$ be a pair of σ-finite measure spaces and let $p_1, p_2 \in [1, \infty)$. Given a measurable function f on $(E_1 \times E_2, \mathcal{B}_1 \times \mathcal{B}_2)$, define

$$\|f\|_{L^{(p_1,p_2)}(\mu_1,\mu_2)} \equiv \left[\int_{E_2} \left(\int_{E_1} |f(x_1, x_2)|^{p_1} \, \mu_1(dx_1) \right)^{p_2/p_1} \mu_2(dx_2) \right]^{1/p_2},$$

and let $L^{(p_1,p_2)}(\mu_1, \mu_2)$ denote the space of $\mathbb{R}$-valued, $\mathcal{B}_1 \times \mathcal{B}_2$-measurable f's for which $\|f\|_{L^{(p_1,p_2)}(\mu_1,\mu_2)} < \infty$. Obviously, when $p_1 = p = p_2$, $\|f\|_{L^{(p_1,p_2)}(\mu_1,\mu_2)} = \|f\|_{L^p(\mu_1 \times \mu_2)}$ and $L^{(p_1,p_2)}(\mu_1, \mu_2) = L^p(\mu_1 \times \mu_2)$.

V.2.11 Lemma. *For all f and g which are measurable on $(E_1 \times E_2, \mathcal{B}_1 \times \mathcal{B}_2)$ and all $\alpha, \beta \in \mathbb{R}$,*

$$\|\alpha f + \beta g\|_{L^{(p_1,p_2)}(\mu_1,\mu_2)} \leq |\alpha| \, \|f\|_{L^{(p_1,p_2)}(\mu_1,\mu_2)} + |\beta| \, \|g\|_{L^{(p_1,p_2)}(\mu_1,\mu_2)}$$

(V.2.12)

$$\|f \cdot g\|_{L^1(\mu_1 \times \mu_2)} \leq \|f\|_{L^{(p_1,p_2)}(\mu_1,\mu_2)} \|g\|_{L^{(p_1',p_2')}(\mu_1,\mu_2)}.$$

Moreover, if $\{f_n\}_1^\infty \cup \{f\} \subseteq L^{(p_1,p_2)}(\mu_1,\mu_2)$, $f_n \longrightarrow f$ (a.e., $\mu_1 \times \mu_2$), and $|f_n| \leq g$ (a.e., $\mu_1 \times \mu_2$) for each $n \geq 1$ and some measurable $g \in L^{(p_1,p_2)}(\mu_1,\mu_2)$, then $\|f_n - f\|_{L^{(p_1,p_2)}(\mu_1,\mu_2)} \longrightarrow 0$. Finally, if μ_1 and μ_2 are finite and $\mathcal{G}$ denotes the class of all ψ's on $E_1 \times E_2$ having the form $\sum_{m=1}^n \chi_{\Gamma_{1,m}}(\cdot_1) \cdot \phi_m(\cdot_2)$ for some $n \geq 1$, $\{\phi_m\}_1^n \subseteq L^\infty(\mu_2)$, and mutually disjoint $\Gamma_{1,1}, \ldots, \Gamma_{1,n} \in \mathcal{B}_1$, then for every measurable $f \in L^{(p_1,p_2)}(\mu_1,\mu_2)$ and $\epsilon > 0$ there is a $\psi \in \mathcal{G}$ such that $\|f - \psi\|_{L^{(p_1,p_2)}(\mu_1,\mu_2)} < \epsilon$.

PROOF: Note that

$$(V.2.13) \qquad \|f\|_{L^{(p_1,p_2)}(\mu_1,\mu_2)} = \big\| \, \|f(\cdot_1,\cdot_2)\|_{L^{p_1}(\mu_1)} \, \big\|_{L^{p_2}(\mu_2)}.$$

Hence the assertions in (V.2.12) are consequences of repeated application of MINKOWSKI's and HÖLDER's inequalities, respectively. Moreover, to prove the second statement, observe (cf. Exercise III.4.11) that for μ_2-almost every $x_2 \in E_2$, $f_n(\cdot,x_2) \longrightarrow f(\cdot,x_2)$ (a.e., μ_1), $|f_n(\cdot,x_2)| \leq g(\cdot,x_2)$ (a.e., μ_1), and $g(\cdot,x_2) \in L^{p_1}(\mu_1)$. Thus, by part ii) of Theorem V.2.2,

$$\|f_n(\cdot,x_2) - f(\cdot,x_2)\|_{L^{p_1}(\mu_1)} \longrightarrow 0$$

for μ_2-almost every $x_2 \in E_2$. In addition,

$$\|f_n(\cdot,x_2) - f(\cdot,x_2)\|_{L^{p_1}(\mu_1)} \leq 2\|g(\cdot,x_2)\|_{L^{p_1}(\mu_1)}$$

for μ_2-almost every $x_2 \in E_2$ and, by (V.2.13) with g replacing f,

$$\big\| \, \|g(\cdot_1,\cdot_2)\|_{L^{p_1}(\mu_1)} \, \big\|_{L^{p_2}(\mu_2)} < \infty.$$

Hence the required result follows after a second application of ii) in Theorem V.2.2.

We now turn to the final part of the lemma, in which the measures μ_1 and μ_2 are assumed to be finite. Without loss in generality, we will assume that μ_1 and μ_2 are, in fact, probability measures. Because $\mu_1 \times \mu_2$ is also a probability measure, JENSEN's inequality and (V.2.13) imply that

$$\|f - \psi\|_{L^{(p_1,p_2)}(\mu_1,\mu_2)} \leq \|f - \psi\|_{L^q(\mu_1 \times \mu_2)} \quad \text{where} \quad q = p_1 \vee p_2.$$

Hence, it suffices to show that for every bounded measurable f on $(E_1 \times E_2, \mathcal{B}_1 \times \mathcal{B}_2)$ and $\epsilon > 0$ there is a $\psi \in \mathcal{G}$ for which $\|f - \psi\|_{L^q(\mu_1 \times \mu_2)} < \epsilon$. But, by part iv) of Theorem V.2.2, the class of simple functions having the form

$$\psi = \sum_{m=1}^n a_m \chi_{\Gamma_{1,m} \times \Gamma_{2,m}}$$

with $\Gamma_{i,m} \in \mathcal{B}_i$ is dense in $L^q(\mu_1 \times \mu_2)$. Thus we will be done once we check that such a ψ is an element of $\mathcal{G}$. To this end, we use the same technique as we did in the final part of the proof of Lemma III.2.3. That is, set $\mathcal{I} = (\{0,1\})^n$ and, for $\boldsymbol{\eta} \in \mathcal{I}$, define $\Gamma_{1,\boldsymbol{\eta}} = \bigcap_{m=1}^n \Gamma_{1,m}^{(\eta_m)}$ where $\Gamma^{(0)} \equiv \Gamma^\complement$ and $\Gamma^{(1)} \equiv \Gamma$. Then

$$\psi(x_1, x_2) = \sum_{m=1}^n a_m \left(\sum_{\boldsymbol{\eta} \in \mathcal{I}} \eta_m \chi_{\Gamma_{1,\boldsymbol{\eta}}}(x_1) \right) \chi_{\Gamma_{2,m}}(x_2) = \sum_{\boldsymbol{\eta} \in \mathcal{I}} \chi_{\Gamma_{1,\boldsymbol{\eta}}}(x_1) \, \phi_{\boldsymbol{\eta}}(x_2),$$

where

$$\phi_{\boldsymbol{\eta}} = \sum_{m=1}^n \eta_m a_m \chi_{\Gamma_{2,m}}.$$

Since the $\Gamma_{1,\boldsymbol{\eta}}$'s are mutually disjoint, this completes the proof. ∎

For our purposes, the most important fact that comes out of these considerations is the following **continuous version of MINKOWSKI's inequality**.

V.2.14 Theorem. *Let $(E_i, \mathcal{B}_i, \mu_i)$, $i \in \{1,2\}$, be σ-finite measure spaces. Then for any $1 \leq p_1 \leq p_2 < \infty$ and any measurable function f on $(E_1 \times E_2, \mathcal{B}_1 \times \mathcal{B}_2)$, $\|f\|_{L^{(p_1,p_2)}(\mu_1,\mu_2)} \leq \|f\|_{L^{(p_2,p_1)}(\mu_2,\mu_1)}$.*

PROOF: Since it is easy to reduce the general case to the one in which both μ_1 and μ_2 are finite, we may take them to be finite. In fact, without loss in generality, we will assume from the outset that they are probability measures.

Let $\mathcal{G}$ be the class described in the last part of Lemma V.2.11. Given $\psi = \sum_1^n \chi_{\Gamma_{1,m}}(\cdot_1) \phi_m(\cdot_2)$ which is an element of $\mathcal{G}$, note that, since the $\Gamma_{1,m}$'s are mutually disjoint, $|\sum_1^n a_m \chi_{\Gamma_{1,m}}|^r = \sum_1^n |a_m|^r \chi_{\Gamma_{1,m}}$ for any $r \in [0,\infty)$ and $a_1, \ldots, a_n \in \mathbb{R}$. Hence, by MINKOWSKI's inequality for $p = p_2/p_1$,

$$\|\psi\|_{L^{(p_1,p_2)}(\mu_1,\mu_2)} = \left[\int_{E_2} \left(\sum_1^n \mu_1(\Gamma_{1,m}) |\phi_m(x_2)|^{p_1} \right)^{p_2/p_1} \mu_2(dx_2) \right]^{1/p_2}$$

$$= \left\| \sum_1^n \mu_1(\Gamma_{1,m}) |\phi_m(\cdot_2)|^{p_1} \right\|_{L^{p_2/p_1}(\mu_2)}^{1/p_1}$$

$$\leq \left[\sum_1^n \mu_1(\Gamma_{1,m}) \big\| |\phi_m|^{p_1} \big\|_{L^{p_2/p_1}(\mu_2)} \right]^{1/p_1}$$

$$= \left[\sum_1^n \mu_1(\Gamma_{1,m}) \|\phi_m\|_{L^{p_2}(\mu_2)}^{p_1} \right]^{1/p_1}$$

$$= \left[\int_{E_1} \sum_1^n \chi_{\Gamma_{1,m}}(x_1) \|\phi_m\|_{L^{p_2}(\mu_2)}^{p_1} \, \mu_1(dx_1) \right]^{1/p_1}$$

$$= \left[\int_{E_1} \left(\sum_1^n \chi_{\Gamma_{1,m}}(x_1) \|\phi_m\|_{L^{p_2}(\mu_2)}^{p_2} \right)^{p_1/p_2} \mu_1(dx_1) \right]^{1/p_1}$$

$$= \left[\int_{E_1} \left(\int_{E_2} \left| \sum_1^n \chi_{\Gamma_{1,m}}(x_1) \phi_m(x_2) \right|^{p_2} \mu_2(dx_2) \right)^{p_1/p_2} \mu_1(dx_1) \right]^{1/p_1}$$

$$= \|\psi\|_{L^{(p_2,p_1)}(\mu_2,\mu_1)}.$$

Hence, we are done when the function f is an element of $\mathcal{G}$.

To complete the proof, let f be a measurable function on $(E_1 \times E_2, \mathcal{B}_1 \times \mathcal{B}_2)$. Clearly we may assume that $\|f\|_{L^{(p_2,p_1)}(\mu_2,\mu_1)} < \infty$. Using the last part of Lemma V.2.11, choose $\{\psi_n\}_1^\infty \subseteq \mathcal{G}$ so that $\|\psi_n - f\|_{L^{(p_2,p_1)}(\mu_2,\mu_1)} \longrightarrow 0$. Then, by JENSEN's inequality, it is easy to check that $\|\psi_n - f\|_{L^1(\mu_1 \times \mu_2)} \longrightarrow 0$, and therefore that $\psi_n \longrightarrow f$ in $\mu_1 \times \mu_2$-measure. Hence, without loss in generality, we will assume that $\psi_n \longrightarrow f$ (a.e., $\mu_1 \times \mu_2$). In particular, by FATOU's Lemma and Exercise III.4.11, this means that

$$\int_{E_1} |f(x_1, x_2)|^{p_1} \, d\mu_1(dx_1) \leq \varliminf_{n \to \infty} \int_{E_1} |\psi_n(x_1, x_2)|^{p_1} \, \mu_1(dx_1)$$

for μ_2-almost every $x_2 \in E_2$. Hence, by the result for $\mathcal{G}$ and another application of FATOU's Lemma, the required result follows for f. ∎

The following result is typical of the way in which one applies Theorem V.2.14.

V.2.15 Theorem. *Let $(E_1, \mathcal{B}_1, \mu_1)$ and $(E_2, \mathcal{B}_2, \mu_2)$ be a pair of σ-finite measure spaces, and suppose that K is a measurable function on $(E_1 \times E_2, \mathcal{B}_1 \times \mathcal{B}_2)$ which satisfies*

$$M_1 \equiv \sup_{x_2 \in E_2} \|K(\cdot, x_2)\|_{L^q(\mu_1)} < \infty \quad \text{and} \quad M_2 \equiv \sup_{x_1 \in E_1} \|K(x_1, \cdot)\|_{L^q(\mu_2)} < \infty$$

for some $q \in [1, \infty)$. Define

$$(V.2.16) \qquad \mathcal{K}f(x_1) = \int_{E_2} K(x_1, x_2) f(x_2) \, \mu_2(dx_2)$$

for $f \in L^{q'}(\mu_2)$. *Then for all $p \in [1, \infty]$ satisfying $\frac{1}{r} \equiv \frac{1}{p} + \frac{1}{q} - 1 \geq 0$,*

$$(\mathrm{V}.2.17) \qquad \|\mathcal{K}f\|_{L^r(\mu_1)} \leq M_1^{q/r} M_2^{1-q/r} \|f\|_{L^p(\mu_2)}.$$

PROOF: First suppose that $r = \infty$ and therefore that $p = q'$. Then, by part i) of Theorem V.2.7,

$$|\mathcal{K}f(x_1)| \leq \|K(x_1, \cdot)\|_{L^q(\mu_2)} \|f\|_{L^p(\mu_2)} \quad \text{for all} \quad x_1 \in E_1;$$

and so (V.2.17) is trivial in this case.

Next, suppose that $p = 1$ and therefore that $q = r$. Noting that $\|\mathcal{K}f\|_{L^r(\mu_1)}$ $\leq \|K \cdot f\|_{L^{(1,r)}(\mu_2, \mu_1)}$, we can apply Theorem V.1.14 to obtain

$$\begin{aligned}
\|\mathcal{K}f\|_{L^r(\mu_1)} &\leq \|K \cdot f\|_{L^{(r,1)}(\mu_1, \mu_2)} \\
&= \int_{E_2} \left(\int_{E_1} |K(x_1, x_2) f(x_2)|^r \, \mu_1(dx_1) \right)^{1/r} \mu_2(dx) \\
&= \int_{E_2} \|K(\cdot, x_2)\|_{L^r(\mu_1)} |f(x_2)| \, \mu_2(dx_2) \leq M_1 \|f\|_{L^1(\mu_2)}.
\end{aligned}$$

Finally, the only case remaining is when $r \in [1, \infty)$ and $p \in (1, \infty)$. Noting that $r \in (q, \infty)$, set $\alpha = q/r$. Then, $\alpha \in (0, 1)$ and $(1 - \alpha)p' = q$. Given $g \in L^{r'}(\mu_1)$, we have, by the second inequality in (V.2.12), that

$$\begin{aligned}
\|g \cdot \mathcal{K}f\|_{L^1(\mu_1)} &\leq \|g \cdot K \cdot f\|_{L^1(\mu_1 \times \mu_2)} \\
&\leq \big\| |K|^\alpha \cdot f \big\|_{L^{(r,p)}(\mu_1, \mu_2)} \big\| g \cdot |K|^{1-\alpha} \big\|_{L^{(r', p')}(\mu_1, \mu_2)}.
\end{aligned}$$

Next, observe that

$$\big\| |K|^\alpha \cdot f \big\|_{L^{(r,p)}(\mu_1, \mu_2)}$$

$$= \left[\int_{E_2} \left(\int_{E_1} |K(x_1, x_2)|^{\alpha r} |f(x_2)|^r \, \mu_1(dx_1) \right)^{p/r} \mu_2(dx_2) \right]^{1/p} \leq M_1^\alpha \|f\|_{L^p(\mu_2)}.$$

At the same time, since $p \leq r$ and therefore $r' \leq p'$, we can apply Theorem V.2.14 to see that $\|g \cdot |K|^{1-\alpha}\|_{L^{(r', p')}(\mu_1, \mu_2)} \leq \|g \cdot |K|^{1-\alpha}\|_{L^{(p', r')}(\mu_2, \mu_1)}$. Hence, by the same reasoning as we just applied to $\||K|^\alpha \cdot f\|_{L^{(r,p)}(\mu_1, \mu_2)}$, we find that $\|g \cdot |K|^{1-\alpha}\|_{L^{(r', p')}(\mu_1, \mu_2)} \leq M_2^{1-\alpha} \|g\|_{L^{r'}(\mu_2)}$. Combining these two, we arrive at $\|g \cdot \mathcal{K}f\|_{L^1(\mu_1)} \leq M_1^\alpha M_2^{1-\alpha} \|f\|_{L^p(\mu_1)}$ for all $g \in L^{r'}(\mu_1)$ with $\|g\|_{L^{r'}(\mu_1)} \leq 1$; and so (V.2.17) now follows from part iii) of Theorem V.2.7. ∎

V.2.18 Corollary. *Let everything be as in Theorem V.2.15, and, for measurable $f : E_2 \longrightarrow \mathbb{R}$, define*

$$\Lambda_K(f) = \left\{ x_1 \in E_1 : \int_{E_2} |K(x_1, x_2)| \, |f(x_2)| \, \mu_2(dx_2) < \infty \right\}$$

and

$$\overline{\mathcal{K}}f(x_1) = \begin{cases} \int_{E_2} K(x_1, x_2) \, f(x_2) \, \mu_2(dx_2) & \text{if } x_1 \in \Lambda_K(f) \\ 0 & \text{otherwise.} \end{cases}$$

Next, let $p \in [1, \infty]$ satisfying $\frac{1}{p} + \frac{1}{q} \geq 1$ be given and define $r \in [1, \infty]$ by $\frac{1}{r} = \frac{1}{p} + \frac{1}{q} - 1$. Then

$$(V.2.19) \qquad \mu_1 \left(\Lambda_K(f)^{\complement} \right) = 0 \quad \text{and} \quad \|\overline{\mathcal{K}}f\|_{L^r(\mu_1)} \leq M_1^{q/r} M_2^{1 - q/r} \|f\|_{L^p(\mu_2)}$$

for $f \in L^p(\mu_2)$. In particular, $\overline{\mathcal{K}}$ maps $L^p(\mu_2)$ linearly into $L^r(\mu_1)$. In fact, $f \in L^p(\mu_2) \longmapsto \overline{\mathcal{K}}f \in L^r(\mu_1)$ is the unique continuous mapping from $L^p(\mu_2)$ into $L^r(\mu_1)$ whose restriction to $L^p(\mu_2) \cap L^{q'}(\mu_2)$ is given by the map $\mathcal{K}$ in (V.2.16).

PROOF: If $r = \infty$, and therefore $p = q'$, there is nothing to do. Thus, we will assume that r and therefore p are finite.

Let $f \in L^p(\mu_2)$ be given, and set $f_n = f \chi_{[-n, n]} \circ f$ for $n \in \mathbb{Z}^+$. Because $p < q'$ and $f_n \in L^p(\mu_2) \cap L^\infty(\mu_2)$, $f \in L^p(\mu_2) \cap L^{q'}(\mu_2)$. Hence, by (V.2.17) applied to $|K|$ and $|f_n|$,

$$\int_{E_1} \left(\int_{\{x_2 : |f(x_2)| \leq n\}} |K(x_1, x_2)| \, |f_n(x_2)| \, \mu_2(dx_2) \right)^r \mu_1(dx_1)$$

$$\leq M_1^q M_2^{r-q} \|f_n\|_{L^p(\mu_2)}^r \leq M_1^q M_2^{r-q} \|f\|_{L^p(\mu_2)}^r.$$

In particular, by FATOU's Lemma, this proves both parts of (V.2.19). Furthermore, if $f, g \in L^p(\mu_2)$ and $\alpha, \beta \in \mathbb{R}$, then

$$\overline{\mathcal{K}}(\alpha f + \beta g) = \alpha \overline{\mathcal{K}}f + \beta \overline{\mathcal{K}}g \quad \text{on} \quad \Lambda_K(f) \cap \Lambda_K(g).$$

Thus, since both $\Lambda_K(f)^{\complement}$ and $\Lambda_K(g)^{\complement}$ have μ_1-measure 0, we now see that, as a mapping into $L^r(\mu_1)$, $\overline{\mathcal{K}}$ is linear. Finally, it is obvious that $\overline{\mathcal{K}}f = \mathcal{K}f$ for $f \in L^{q'}(\mu_2)$. Hence, if $\mathcal{K}'$ is any extension of $\mathcal{K}$ on $L^p(\mu_2) \cap L^{q'}(\mu_2)$ as a continuous mapping from $L^p(\mu_2)$ to $L^r(\mu_1)$, then (with the same choice of $\{f_n\}_1^\infty$ as above)

$$\left\|\overline{\mathcal{K}}f - \mathcal{K}'f\right\|_{L^r(\mu_1)} \leq \varlimsup_{n \to \infty} \left\|\overline{\mathcal{K}}f - \overline{\mathcal{K}}f_n\right\|_{L^r(\mu_1)}$$

$$= \varlimsup_{n \to \infty} \left\|\overline{\mathcal{K}}(f - f_n)\right\|_{L^r(\mu_1)} \leq M_1^{\frac{a}{r}} M_2^{1-\frac{a}{r}} \varlimsup_{n \to \infty} \|f - f_n\|_{L^p(\mu_1)} = 0. \quad \blacksquare$$

V.2.20 Exercise.

Let $(E, \mathcal{B}, \mu)$ be a measure space and let $1 \leq q_1 \leq q_2 \leq \infty$ be given. If $f \in L^{q_1}(\mu) \cap L^{q_2}(\mu)$, show that for any $t \in (0, 1)$

$$\text{(V.2.21)} \qquad \|f\|_{L^{p_t}(\mu)} \leq \|f\|_{L^{q_1}(\mu)}^t \|f\|_{L^{q_2}(\mu)}^{1-t} \qquad \text{where } \frac{1}{p_t} = \frac{t}{q_1} + \frac{1-t}{q_2}.$$

Note that (IV.2.21) says that $p \longmapsto -\log \|f\|_{L^p(\mu)}$ is a concave function of $1/p$.

V.2.22 Exercise.

i) If $(E, \mathcal{B}, \mu)$ is a probability space, show that $p \in [1, \infty] \longmapsto \|f\|_{L^p(\mu)}$ is a non-decreasing function for any measurable f on $(E, \mathcal{B})$.

ii) Let $E = \mathbb{Z}^+$ and define μ on $\mathcal{B} = \mathcal{P}(E)$ by $\mu(\{n\}) = 1$ for all $n \in \mathbb{Z}^+$. In this case show that $p \in [1, \infty] \longmapsto \|f\|_{L^p(\mu)}$ is non-increasing for every f on E.

iii) Assuming that $(E, \mathcal{B}, \mu)$ is a finite measure space, show that $\|f\|_{L^\infty(\mu)} = \lim_{p \to \infty} \|f\|_{L^p(\mu)}$ for every measurable function f on $(E, \mathcal{B})$. Conclude from this that if $(E_i, \mathcal{B}_i)$, $i \in \{1, 2\}$, are measurable spaces and μ_2 is a σ-finite measure on $(E_2, \mathcal{B}_2)$, then for every measurable function f on $(E_1 \times E_2, \mu_1 \times \mu_2)$ the function $x_1 \longmapsto \|f(x_1, \cdot)\|_{L^\infty(\mu_2)}$ is measurable on $(E_1, \mathcal{B}_1)$. In particular, we could have defined $\|f\|_{L^{(p_1, p_2)}(\mu_1, \mu_2)}$ for all $p_1, p_2 \in [1, \infty]$.

iv) Show that if $(E, \mathcal{B}, \mu)$ is any measure space and $f \in L^1(\mu)$, then $\|f\|_{L^\infty(\mu)} = \lim_{p \to \infty} \|f\|_{L^p(\mu)}$.

V.3. Convolution and Approximate Identities.

We will use the ideas of the last section to develop in this section an important notion of *multiplication* for functions on $\mathbb{R}^N$; and, because the only measure involved will be LEBESGUE's, we will use the notation $L^p(\mathbb{R}^N)$ instead of the more cumbersome $L^p(\lambda_{\mathbb{R}^N})$.

V.3.1 Theorem. (YOUNG'S INEQUALITY) *Let p and q from $[1, \infty]$ satisfying $\frac{1}{p} + \frac{1}{q} \geq 1$ be given, and define $r \in [1, \infty]$ by $\frac{1}{r} = \frac{1}{p} + \frac{1}{q} - 1$. Then, for each $f \in L^p(\mathbb{R}^N)$ and $g \in L^q(\mathbb{R}^N)$, the complement of the set*

$$(V.3.2) \qquad \Lambda(f, g) \equiv \left\{ x \in \mathbb{R}^N : \int_{\mathbb{R}^N} |f(x-y)|\, |g(y)|\, dy < \infty \right\}$$

has LEBESGUE measure 0. Furthermore, if

$$(V.3.3) \qquad f * g(x) \equiv \begin{cases} \int_{\mathbb{R}^N} f(x-y)\, g(y)\, dy & \text{when} \quad x \in \Lambda(f, g) \\ 0 & \text{otherwise,} \end{cases}$$

then $f * g = g * f$ and

$$(V.3.4) \qquad \|f * g\|_{L^r(\mathbb{R}^N)} \leq \|f\|_{L^p(\mathbb{R}^N)} \|g\|_{L^q(\mathbb{R}^N)}.$$

Finally, the mapping $(f, g) \in L^p(\mathbb{R}^N) \times L^q(\mathbb{R}^N) \longmapsto f * g \in L^r(\mathbb{R}^N)$ is bilinear.

PROOF: We begin with the observation that there is nothing to do when $r = \infty$. Thus, we will assume throughout that r and therefore also p and q are all finite. Next, using the translation invariance of LEBESGUE's measure, first note that $\Lambda(f, g) = \Lambda(g, f)$ and then conclude that $f * g = g * f$. Finally, given $q \in [1, \infty)$ and $g \in L^q(\mathbb{R}^N)$, set $K(x, y) = g(x - y)$ for x, $y \in \mathbb{R}^N$. Obviously,

$$\sup_{y \in \mathbb{R}^N} \|K(\cdot, y)\|_{L^q(\mathbb{R}^N)} = \sup_{x \in \mathbb{R}^N} \|K(x, \cdot)\|_{L^q(\mathbb{R}^N)} = \|g\|_{L^q(\mathbb{R}^N)} < \infty;$$

and, in the notation of Corollary V.2.18, $\Lambda(f, g) = \Lambda_K(f)$ and $f * g = \overline{K}f$. In particular, for each $f \in L^p(\mathbb{R}^N)$, $\Lambda(f, g)^\complement$ has LEBESGUE measure 0 and (V.3.4) holds. In addition, $f \in L^p(\mathbb{R}^N) \longmapsto f * g \in L^r(\mathbb{R}^N)$ is linear for each $g \in L^q(\mathbb{R}^N)$; and therefore the bilinearity assertion follows after one reverses the roles of f and g. $\blacksquare$

The quantity $f * g$ described in (V.3.3) is called the **convolution of f times g**. In applications, the most useful cases are those when $f \in L^p(\mathbb{R}^N)$ and $g \in L^q(\mathbb{R}^N)$ where either $p = q'$ (and therefore $r = \infty$) or $p = 1$ (and therefore $r = q$). To get more information about the case when $p = q'$, we will need the following.

V.3.5 Lemma. *Given $h \in \mathbb{R}^N$, define $\tau_h f$ for functions f on $\mathbb{R}^N$ by $\tau_h f(x) = f(x + h)$. Then τ_h is an isometry on $L^p(\mathbb{R}^N)$ for every $h \in \mathbb{R}^N$ and $p \in [1, \infty]$. Moreover, if $p \in [1, \infty)$ and $f \in L^p(\mathbb{R}^N)$, then*

$$(V.3.6) \qquad \lim_{h \to 0} \|\tau_h f - f\|_{L^p(\mathbb{R}^N)} = 0.$$

PROOF: The first assertion is an immediate consequence of the translation invariance of LEBESGUE's measure.

Next, suppose that $p \in [1, \infty)$ is given. If $\mathcal{G}$ denotes the class of $f \in L^p(\mathbb{R}^N)$ for which (V.3.6) holds, it is clear that $C_c(\mathbb{R}^N) \subseteq \mathcal{G}$. Hence, by **v**) in Theorem V.2.2, we will know that $\mathcal{G} = L^p(\mathbb{R}^N)$ as soon as we show that $\mathcal{G}$ is closed in $L^p(\mathbb{R}^N)$. To this end, let $\{f_n\}_1^\infty \subseteq \mathcal{G}$ and suppose that $f_n \longrightarrow f$ in $L^p(\mathbb{R}^N)$. Then

$$
\overline{\lim_{h \to 0}} \, \|\tau_h f - f\|_{L^p(\mathbb{R}^N)}
$$

$$
\leq \overline{\lim_{h \to 0}} \, \|\tau_h(f - f_n)\|_{L^p(\mathbb{R}^N)} + \overline{\lim_{h \to 0}} \, \|\tau_h f_n - f_n\|_{L^p(\mathbb{R}^N)} + \|f_n - f\|_{L^p(\mathbb{R}^N)}
$$

$$
= 2\|f_n - f\|_{L^p(\mathbb{R}^N)} \longrightarrow 0
$$

as $n \to \infty$. ∎

V.3.7 Theorem. *Let $p \in [1, \infty]$, $f \in L^p(\mathbb{R}^N)$, and $g \in L^{p'}(\mathbb{R}^N)$. Then*

$$
\tau_h(f * g) = (\tau_h f) * g = f * (\tau_h g) \quad \text{for all} \quad h \in \mathbb{R}^N.
$$

Moreover, $f * g$ is uniformly continuous on $\mathbb{R}^N$ and

(V.3.8) $\|f * g\|_u \leq \|f\|_{L^p(\mathbb{R}^N)} \|g\|_{L^{p'}(\mathbb{R}^N)}.$

Finally, if $p \in (1, \infty)$, then

(V.3.9) $\displaystyle \lim_{|x| \to \infty} f * g(x) = 0.$

PROOF: The first assertion is again just an expression of translation invariance for $\lambda_{\mathbb{R}^N}$. Further, (V.3.8) is guaranteed by part **i**) of Theorem V.2.7. To see that $f * g$ is uniformly continuous, first suppose that $p \in [1, \infty)$. Then, by (V.3.8) and (V.3.6),

$$
\|\tau_h(f * g) - f * g\|_u = \|(\tau_h f - f) * g\|_u \leq \|\tau_h f - f\|_{L^p(\mathbb{R}^N)} \|g\|_{L^{p'}(\mathbb{R}^N)} \longrightarrow 0
$$

as $|h| \to 0$; and when $p = \infty$, simply reverse the roles of f and g in this argument.

To prove the final assertion, first let $f \in C_c(\mathbb{R}^N)$ be given, and define $\mathcal{G}_f$ be the class of $g \in L^{p'}(\mathbb{R}^N)$ for which (V.3.9) holds. Then it is easy to check that $C_c(\mathbb{R}^N) \subseteq \mathcal{G}_f$. Moreover, by (V.3.8), one sees that $\mathcal{G}_f$ is closed in $L^{p'}(\mathbb{R}^N)$. Hence, just as in the final step of the proof of Lemma V.3.5, we conclude that $\mathcal{G}_f = L^{p'}(\mathbb{R}^N)$. Next, let $g \in L^{p'}(\mathbb{R}^N)$ be given and define $\mathcal{H}_g$ to be the class of $f \in L^p(\mathbb{R}^N)$ for which (V.3.9) is true. By the preceding, we know that $C_c(\mathbb{R}^N) \subseteq \mathcal{H}_g$. Moreover, just as before, $\mathcal{H}_g$ is closed in $L^p(\mathbb{R}^N)$; and therefore $\mathcal{H}_g = L^p(\mathbb{R}^N)$. ∎

V.3.10 Remark.

Both Theorem V.3.3 and Theorem V.3.7 tell us that the convolution of two functions is often more regular than either or both of its factors. An application of this fact is given in Exercise V.3.26 below, where one sees how it leads to an elegant derivation of Lemma II.1.15.

The next result can be considered as another example of the observation made in the preceding Remark.

V.3.11 Lemma. *Let $g \in C^1(\mathbb{R}^N)$, and assume that g as well as $g_{,1}, \ldots, g_{,N}$ are elements of $L^{p'}(\mathbb{R}^N)$ for some $p \in [1, \infty]$. (Recall that $g_{,i} \equiv \frac{\partial g}{\partial x_i}$.) Then $f * g \in C^1(\mathbb{R}^N)$ for every $f \in L^p(\mathbb{R}^N)$, and*

$$(V.3.12) \qquad\qquad \frac{\partial f * g}{\partial x_i} = f * g_{,i}, \quad 1 \leq i \leq N.$$

PROOF: Let $\omega \in S^{N-1}$ be given. If $p' \in [1, \infty)$, then, by Theorem V.3.7, $f * g(\cdot + t\omega) - f * g = f * (\tau_{t\omega}g - g)$ for every $t \in \mathbb{R}$. Since

$$\frac{\tau_{t\omega}g(y) - g(y)}{t} = \int_{[0,1]} (\omega, \nabla g(y + st\omega))_{\mathbb{R}^N} \, ds$$

and, by Theorem V.2.14,

$$\left\| \int_{[0,1]} (\omega, \nabla g(\cdot + st\omega) - \nabla g(\cdot))_{\mathbb{R}^N} \, ds \right\|_{L^{p'}(\mathbb{R}^N)}$$

$$\leq \int_{[0,1]} \left\| \tau_{st\omega}(\omega, \nabla g)_{\mathbb{R}^N} - (\omega, \nabla g)_{\mathbb{R}^N} \right\|_{L^{p'}(\mathbb{R}^N)} \, ds \longrightarrow 0$$

as $t \to 0$, the required result follows from (V.3.4). On the other hand, if $p' = \infty$, then

$$\frac{\tau_{t\omega}g(y) - g(y)}{t} = \int_{[0,1]} (\omega, \nabla g(y + st\omega))_{\mathbb{R}^N} \, ds \longrightarrow (\omega, \nabla g(y))_{\mathbb{R}^N}$$

boundedly and point-wise, and therefore the result follows, in this case, from LEBESGUE's Dominated Convergence Theorem. ∎

The preceding result leads immediately to the conclusion that the smoother g is the smoother is $f * g$. More precisely, given a multi-index $\alpha = (\alpha_1, \ldots, \alpha_N)$, where the α_i's are non-negative integers, define $|\alpha| = \sum_1^N \alpha_i$ and

$$\partial^\alpha = \frac{\partial^{|\alpha|}}{\partial x^\alpha} \equiv \frac{\partial^{\alpha_1}}{\partial x_1^{\alpha_1}} \cdots \frac{\partial^{\alpha_N}}{\partial x_1^{\alpha_N}} (\equiv \text{Identity when } |\alpha| = 0).$$

Then, as an immediate corollary of Lemma V.3.11, we see that if $g \in C^{\infty}(\mathbb{R}^N)$ and $\partial^{\alpha} g \in L^{p'}(\mathbb{R}^N)$ for some $p \in [1, \infty]$ and all α's, then

$$(V.3.13) \qquad f * g \in C^{\infty}(\mathbb{R}^N) \quad \text{and} \quad \partial^{\alpha}(f * g) = f * (\partial^{\alpha} g)$$

for every $f \in L^p(\mathbb{R}^N)$.

We next turn our attention to the case when $g \in L^1(\mathbb{R}^N)$. The main result here is the one which follows.

V.3.14 Theorem. *Given $g \in L^1(\mathbb{R}^N)$ and $t > 0$, define $g_t(\cdot) = t^{-N} g(\cdot/t)$. Then $g_t \in L^1(\mathbb{R}^N)$ and $\int g_t\, dx = \int g\, dx$. In addition, if $\int g\, dx = 1$, then for every $p \in [1, \infty)$ and $f \in L^p(\mathbb{R}^N)$:*

$$(V.3.15) \qquad \lim_{t \searrow 0} \left\| f * g_t - f \right\|_{L^p(\mathbb{R}^N)} = 0.$$

PROOF: We need only deal with the last statement.

Assume that $\int g\, dx = 1$. Given $f \in L^p(\mathbb{R}^N)$, note that, for almost every $x \in \mathbb{R}^N$,

$$f * g_t(x) - f(x) = \int_{\mathbb{R}^N} \big(f(x-y) - f(x)\big) g_t(y)\, dy = \int_{\mathbb{R}^N} \big(f(x-ty) - f(x)\big) g(y)\, dy.$$

Hence, if $\Psi^t(x, y) = \big(f(x - ty) - f(x)\big) g(y)$, then, by Theorem V.2.14,

$$\left\| f * g_t - f \right\|_{L^p(\mathbb{R}^N)} \le \left\| \Psi^t \right\|_{L^{(1,p)}(\lambda_{\mathbb{R}^N}, \lambda_{\mathbb{R}^N})}$$

$$\le \left\| \Psi^t \right\|_{L^{(p,1)}(\lambda_{\mathbb{R}^N}, \lambda_{\mathbb{R}^N})} = \int_{\mathbb{R}^N} \left\| \tau_{-ty} f - f \right\|_{L^p(\mathbb{R}^N)} |g(y)|\, dy.$$

Since $\left\| \tau_{-ty} f - f \right\|_{L^p(\mathbb{R}^N)} \le 2\|f\|_{L^p(\mathbb{R}^N)}$ and because of (V.3.6), we now see that the result follows from the above by LEBESGUE's Dominated Convergence Theorem. ∎

For reasons which ought to be made clear by Theorem V.3.14, if $g \in L^1(\mathbb{R}^N)$ and $\int g\, dx = 1$, the corresponding family $\{g_t : t > 0\}$ is called an **approximate identity**. To understand how an approximate identity actually carries out *an approximation of the identity*, consider the case when g is non-negative and vanishes off of $B(0, 1)$. Then *the volume under the graph* of g_t continues to be 1 as $t \searrow 0$ while the base of the graph is restricted to $B(0, t)$. Hence, all the *mass* is getting concentrated over the origin.

Combining Theorem V.3.14 and (V.3.13), we get the following important approximation procedure.

V.3.16 Corollary. *Let* $g \in C^\infty(\mathbb{R}^N) \cap L^1(\mathbb{R}^N)$ *with* $\int_{\mathbb{R}^N} g(x)\, dx = 1$ *be given. In addition, let* $p \in [1, \infty)$ *and assume that* $\partial^\alpha g \in L^{p'}(\mathbb{R}^N)$ *for all* $\alpha \in \mathbb{N}^N$. *Then, for each* $f \in L^p(\mathbb{R}^N)$, $f * g_t \longrightarrow f$ *in* $L^p(\mathbb{R}^N)$ *as* $t \searrow 0$; *and, for all* $t > 0$, $f * g_t$ *has bounded, continuous derivatives of all orders and* $\partial^\alpha(f * g_t) = f * (\partial^\alpha g_t)$, $\alpha \in \mathbb{N}^N$.

V.3.17 Exercise.

i) Define the **Gauss kernel** $\gamma(x) = (2\pi)^{-N/2} \exp(-|x|^2/2)$ for $x \in \mathbb{R}^N$. Using the result in part i) Exercise IV.2.6, show that $\int_{\mathbb{R}^N} \gamma(x)\, dx = 1$ and that

$$(V.3.18) \qquad \gamma_{s^{1/2}} * \gamma_{t^{1/2}} = \gamma_{(s+t)^{1/2}} \quad \text{for } s,\, t \in (0, \infty).$$

Clearly (V.3.18) says that the approximate identity $\{\gamma_{t^{1/2}} : t \in (0, \infty)\}$ is a **convolution semigroup** of functions. It is known variously as the **heat flow semigroup** or the **Weierstrass semigroup**.

ii) Define ν on $\mathbb{R}$ by

$$\nu(\xi) = \frac{\chi_{(0,\infty)}(\xi) \cdot e^{-1/\xi}}{\pi^{1/2} \xi^{3/2}}.$$

Show that $\int_{\mathbb{R}} \nu(\xi)\, d\xi = 1$ and that

$$(V.3.19) \qquad \nu_{s^2} * \nu_{t^2} = \nu_{(s+t)^2} \quad \text{for } s,\, t \in (0, \infty).$$

Hint: Note that for $\eta \in (0, \infty)$

$$\nu_{s^2} * \nu_{t^2}(\eta) = \frac{st}{\pi} \int_{(0,\eta)} \frac{1}{(\xi(\eta - \xi))^{3/2}} \exp\left[-\frac{s^2}{\xi} - \frac{t^2}{\eta - \xi}\right] d\xi,$$

try the change of variable $\Phi(\xi) = \frac{\xi}{\eta - \xi}$, and use part **iv)** of Exercise IV.2.6. Thus the family $\{\nu_{t^2} : t > 0\}$ is a convolution semigroup. The function $\nu(\cdot)$ or, more precisely, the measure $\Gamma \longmapsto \int_\Gamma \nu(\xi)\, d\xi$ plays a role in probability theory, where it is called the **one-sided stable law of order** $1/2$.

iii) Using part ii) of Exercise IV.3.19, check that the function P on $\mathbb{R}^N$ given by

$$P(x) = \frac{2}{\omega_N}(1 + |x|^2)^{-(N+1)/2}, \quad x \in \mathbb{R}^N,$$

has LEBESGUE integral 1. Next prove the representation

$$(V.3.20) \qquad P_t(x) = \int_{(0,\infty)} \gamma_{(\xi/2)^{1/2}}(x) \nu_{t^2}(\xi)\, d\xi.$$

Finally, using (V.3.20) together with the preceding parts of this exercise, show that

(V.3.21) $$P_s * P_t = P_{s+t}, \quad s, t \in (0, \infty);$$

and therefore that $\{P_t : t > 0\}$ is a convolution semigroup. This semigroup is known as the **Poisson semigroup** among harmonic analysts and as the **Cauchy semigroup** in probability theory; the representation (V.3.20) is an example of how to obtain one semigroup from another by the method of **subordination**.

V.3.22 Exercise.

Show that if μ is a finite measure on $\mathbb{R}^N$ and $p \in [1, \infty]$, then for all $f \in C_c(\mathbb{R}^N)$ the function $f * \mu$ given by

$$f * \mu(x) = \int_{\mathbb{R}^N} f(x - y) \, \mu(dy), \quad x \in \mathbb{R}^N$$

is continuous and satisfies

(V.3.23) $$\|f * \mu\|_{L^p(\mathbb{R}^N)} \leq \mu(\mathbb{R}^N) \|f\|_{L^p(\mathbb{R}^N)}.$$

Next, use (V.3.23) to show that for each $p \in [1, \infty)$ there is a unique continuous map $\overline{\mathcal{K}}_\mu : L^p(\mathbb{R}^N) \longrightarrow L^p(\mathbb{R}^N)$ such that $\overline{\mathcal{K}}_\mu f = f * \mu$ for $f \in C_c(\mathbb{R}^N)$. Finally, note that (V.3.23) continues to hold when $f * \mu$ is replaced by $\overline{\mathcal{K}}_\mu f$, but that $\overline{K}_\mu f$ need not be continuous for every $f \in L^p(\mathbb{R}^N)$.

V.3.24 Exercise.

i) Set

$$c_N = \left(\int_{B(0,1)} \exp\left[-\frac{1}{1 - |x|^2} \right] dx \right)^{-1}$$

and define

$$\rho(x) = \begin{cases} c_N \exp\left[-\frac{1}{1 - |x|^2} \right] & \text{if } x \in B(0, 1) \\ 0 & \text{if } x \notin B(0, 1) \end{cases}.$$

Show that $\rho \in C^\infty(\mathbb{R}^N)$.

ii) Use the preceding to show that if F is a closed subset of $\mathbb{R}^N$ and G is an open subset satisfying $\text{dist}(F, G^\complement) > 0$, then there is an $\eta \in C^\infty(\mathbb{R}^N)$ such that $\chi_F \leq \eta \leq \chi_G$.

iii) Show that for each pair $p, q \in [1, \infty)$ and every $f \in L^p(\mathbb{R}^N) \cap L^q(\mathbb{R}^N)$ there exists a sequence $\psi_n \in C_0^\infty(\mathbb{R}^N)$ such that $\psi_n \longrightarrow f$ both in $L^p(\mathbb{R}^N)$ and in $L^q(\mathbb{R}^N)$. In particular, $C_c^\infty(\mathbb{R}^N)$ is dense in $L^p(\mathbb{R}^N)$ for every $p \in [1, \infty)$.

iv) Let G be an open subset of $\mathbb{R}^N$. When $N \geq 2$, we showed in Theorem IV.4.12 that every $u \in C^2(G)$ which is harmonic on G satisfies the Mean Value Property (IV.4.13) for balls $B(x, R)$ whose closures are in G. Moreover, as pointed out in the paragraph preceding that theorem, the Mean Value Property is a triviality when $N = 1$. In this exercise you are to prove the converse of the Mean Value Property. Namely, you are to show that if $u \in C(G)$ satisfies (IV.4.13) whenever $\overline{B(x, R)} \subseteq G$, then $u \in C^\infty(G)$ and u is harmonic on G. The proof can be accomplished in two steps. First show that if $\overline{B(x, 2t)} \subseteq G$, then the Mean Value Property implies that

$$u(\xi) = \Big[(\chi_{B(x, 2t)} \cdot u) * \rho_t \Big](\xi) \quad \text{for} \quad \xi \in B(x, t);$$

and conclude from this that $u \in C^\infty(G)$. Second, show that for any $f \in C^\infty(G)$ and $x \in G$,

$$(\mathrm{V}.3.25) \qquad \Delta f(x) = \frac{2}{\Omega_N} \lim_{t \searrow 0} \frac{1}{t^2} \int_{\mathbf{S}^{N-1}} \big(f(x + t\omega) - f(x) \big) \lambda_{\mathbf{S}^{N-1}}(d\omega).$$

Hint: To prove (V.3.25), expand f in a two place TAYLOR expansion around x and use the relations in (IV.2.5).

V.3.26 Exercise.

Let $\Gamma \in \overline{\mathcal{B}}_{\mathbb{R}^N}$ have finite LEBESGUE measure and set $u(x) = \chi_{-\Gamma} * \chi_\Gamma$. Show that $u(x) \leq |\Gamma| \chi_\Delta$, where $\Delta = \Gamma - \Gamma \equiv \{ y - x : x, y \in \Gamma \}$, and that $u(0) = |\Gamma|$. Use these observations to give another proof of Lemma II.1.15.

V.3.27 Exercise.

Define the σ-finite measure μ on $\big((0, \infty), \mathcal{B}_{(0, \infty)} \big)$ by

$$\mu(\Gamma) = \int_\Gamma \frac{1}{x} \, dx \quad \text{for} \quad \Gamma \in \mathcal{B}_{(0, \infty)},$$

and show that μ is **invariant under the multiplicative group** in the sense that

$$\int_{(0, \infty)} f(\alpha x) \, \mu(dx) = \int_{(0, \infty)} f(x) \, \mu(dx), \qquad \alpha \in (0, \infty),$$

and

$$\int_{(0,\infty)} f\left(\frac{1}{x}\right) \mu(dx) = \int_{(0,\infty)} f(x)\,\mu(dx)$$

for every $\mathcal{B}_{(0,\infty)}$-measurable $f : (0,\infty) \longrightarrow [0,\infty]$. Next, for $\mathcal{B}_{(0,\infty)}$-measurable, $\mathbb{R}$-valued functions f and g, set

$$\Lambda_\mu(f,g) = \left\{ x \in (0,\infty) : \int_{(0,\infty)} \left| f\left(\frac{x}{y}\right) \right| |g(y)|\,\mu(dy) < \infty \right\},$$

$$f \bullet g(x) = \begin{cases} \int_{(0,\infty)} f\left(\frac{x}{y}\right) g(y)\,\mu(dy) & \text{when} \quad x \in \Lambda_\mu(f,g) \\ 0 & \text{otherwise,} \end{cases}$$

and show that $f \bullet g = g \bullet f$. In addition, show that if $p,\,q \in [1,\infty]$ satisfy $\frac{1}{r} \equiv \frac{1}{p} + \frac{1}{q} - 1 \geq 0$, then

$$\mu\left(\Lambda(f,g)^\complement\right) = 0 \quad \text{and} \quad \|f \bullet g\|_{L^r(\mu)} \leq \|f\|_{L^p(\mu)} \|g\|_{L^q(\mu)}$$

for all $f \in L^p(\mu)$ and $g \in L^q(\mu)$. Finally, use these considerations to prove the following one of G.H. HARDY's many inequalities:

$$\left[\int_{(0,\infty)} \frac{1}{x^{1+\alpha}} \left(\int_{(0,x)} \phi(y)\,dy \right)^p dx \right]^{\frac{1}{p}} \leq \frac{p}{\alpha} \left(\int_{(0,\infty)} \frac{(y\phi(y))^p}{y^{1+\alpha}}\,dy \right)^{\frac{1}{p}}$$

for all $\alpha \in (0,\infty)$ and all non-negative, $\mathcal{B}_{(0,\infty)}$-measurable ϕ.

Hint: To prove everything except HARDY's inequality, simply repeat the argument used in the proof of YOUNG's Inequality. To prove HARDY's result, take

$$f(x) = \left(\frac{1}{x}\right)^{\frac{\alpha}{p}} \chi_{[1,\infty)}(x) \quad \text{and} \quad g(x) = x^{1-\frac{\alpha}{p}} \phi(x),$$

and use $\|f \bullet g\|_{L^p(\mu)} \leq \|f\|_{L^1(\mu)} \|g\|_{L^p(\mu)}$.

Chapter VI. A Little Abstract Theory

VI.1. An Existence Theorem.

In the Chapter II we constructed LEBESGUE's measure on $\mathbb{R}^N$, and in ensuing chapters we saw how to construct various other measures from a given measure. However, as yet, we have not discussed any general procedure for the construction of measures *ab initio*; and it is the purpose of the present section to provide such a procedure.

The basic idea behind what we will be doing appears to be due to F. RIESZ and entails the reversal, in some sense, of the process by which we went in Chapter III from the existence of a measure to the existence of integrals. That is, we will suppose that we have at hand an *integral* and will attempt to show that it must have come from a measure. Thus, we must first describe what we mean by an *integral*.

Let E be a non-empty set. We will say that a subset **L** of the functions $f : E \longrightarrow \overline{\mathbb{R}}$ is a **lattice** if $f \wedge g$ and $f \vee g$ are both elements of $\in$ **L** whenever f and g are. Given a lattice **L** of $\mathbb{R}$-valued functions, we will say that **L** is a **vector lattice** if the constant function **0** is an element of **L** and **L** is a vector space over $\mathbb{R}$. Note that if **L** is a vector space of $\mathbb{R}$-valued functions on E, then it is a vector lattice if and only if $f^+ \equiv f \vee 0 \in \mathbf{L}$ whenever $f \in \mathbf{L}$. Next, given a vector lattice **L**, we will say that the mapping $I : \mathbf{L} \longrightarrow \mathbb{R}$ is an **integral on L** if

1) I is linear,

2) I is **non-negative** in the sense that $I(f) \geq 0$ for every non-negative $f \in \mathbf{L}$,

3) $I(f_n) \searrow 0$ whenever $\{f_n\}_1^\infty \subseteq \mathbf{L}$ is a non-increasing sequence which tends (point-wise) to 0.

Finally, we will say that the triple $(E, \mathbf{L}, I)$ is an **integration theory** if **L** is a vector lattice of functions $f : E \longrightarrow \mathbb{R}$ and I is an integral on **L**.

VI.1.1 Examples.

i) The basic model on which the preceding definitions are based is the one which comes from the integration theory for a finite space $(E, \mathcal{B}, \mu)$. Indeed, in that case, $\mathbf{L} = L^1(\mu)$ and $I(f) = \int f \, d\mu$.

ii) A second basic source of integration theories is the one which comes from *finitely additive functions on an algebra*. That is, let $\mathcal{A}$ be an algebra of subsets of E and denote by $\mathbf{L}(\mathcal{A})$ the space of simple functions $f : E \longrightarrow \mathbb{R}$ with the property that $\{f = a\} \in \mathcal{A}$ for every $a \in \mathbb{R}$. It is then an easy matter to check that $\mathbf{L}(\mathcal{A})$ is a vector lattice. Now let $\mu : \mathcal{A} \longrightarrow [0, \infty)$ be **finitely additive** in the sense that

$$\mu(\Gamma_1 \cup \Gamma_2) = \mu(\Gamma_1) + \mu(\Gamma_2) \quad \text{for disjoint} \quad \Gamma_1, \Gamma_2 \in \mathcal{A}.$$

Note that, since $\mu(\emptyset) = \mu(\emptyset \cup \emptyset) = 2\mu(\emptyset)$, $\mu(\emptyset)$ must be 0. Also, by proceeding in precisely the same way as we did (via Lemma III.2.3) in the proof of Lemma III.2.4, one can show that

$$f \in \mathbf{L}(\mathcal{A}) \longmapsto I(f) \equiv \sum_{a \in \mathrm{range}(f)} a\,\mu(\{f = a\})$$

is linear and non-negative. Finally, observe that I cannot be an integral unless μ has the property that

(VI.1.2) $$\mu(\Gamma_n) \searrow 0 \quad \text{whenever} \quad \{\Gamma_n\}_1^\infty \subseteq \mathcal{A} \text{ decreases to } \emptyset.$$

On the other hand, if (VI.1.2) holds and $\{f_n\}_1^\infty \subseteq \mathbf{L}(\mathcal{A})$ is a non-increasing sequence which tends point-wise to 0, then for each $\epsilon > 0$

$$\varlimsup_{n \to \infty} I(f_n) \leq \epsilon I(\mathbf{1}) + \sup_{x \in E} |f_1(x)| \varlimsup_{n \to \infty} \mu(\{f_n > \epsilon\}) = \epsilon I(\mathbf{1}).$$

Thus, in this setting, (VI.1.2) is equivalent to I being an integral.

iii) A third important example of an integration theory is provided by the following abstraction of RIEMANN's theory. Namely, let E be a compact topological space, and note that $C(E; \mathbb{R})$ is a vector lattice. Next, suppose that $I : C(E; \mathbb{R}) \longrightarrow \mathbb{R}$ is a linear map which is non-negative. It is then clear that $|I(f)| \leq C\|f\|_{\mathrm{u},E}$, $f \in C(E; \mathbb{R})$, where $C = I(\mathbf{1})$ and $\|f\|_{\mathrm{u},E} \equiv \sup_{x \in E} |f(x)|$ is the **uniform norm** of f on E. In particular, this means that $|I(f_n) - I(f)| \leq C\|f_n - f\|_{\mathrm{u},E} \longrightarrow 0$ if $f_n \longrightarrow f$ uniformly. Thus, to see that I is an integral, all that we have to do is use DINI's Lemma (cf. Lemma VI.1.23 below) which says that $f_n \longrightarrow 0$ uniformly on E if $\{f_n\}_1^\infty \subseteq C(E; \mathbb{R})$ decreases point-wise to 0.

Our main goal will be to show that, at least when $1 \in \mathbf{L}$, every integration theory is the sub-theory of the sort of theory described in i) above. Thus, we must learn how to extract the *measure* μ from the *integral*. At least in case ii) above, it is clear how one might begin such a procedure. Namely, $\mathcal{A} = \{\Gamma \subseteq E : \chi_\Gamma \in \mathbf{L}(\mathcal{A})\}$ and $\mu(\Gamma) = I(\chi_\Gamma)$ for $\Gamma \in \mathcal{A}$. Hence, what we are attempting to do in this case is tantamount to showing that μ can be extended as a measure to the σ-algebra $\sigma(\mathcal{A})$ generated by $\mathcal{A}$. On the other hand, it is not so immediately clear where to start looking for the measure μ in case iii); the procedure which got us started in case ii) does not work here since there will seldom be many $\Gamma \subseteq E$ for which $\chi_\Gamma \in C(E;\mathbb{R})$. Thus, what we must do first is show that I can be extended to a larger class of functions $f : E \longrightarrow \mathbb{R}$ and only then look for μ.

Our extension procedure has two steps, the first of which is nothing but a re-run of what we did in Section III.2 and the second one is a minor variant on what we did in Section II.1.

VI.1.3 Lemma. *Let $(E, \mathbf{L}, I)$ be an integration theory and define $\mathbf{L}_u$ to be the class of $f : E \longrightarrow (-\infty, \infty]$ which can be written as the point-wise limit of a non-decreasing sequence $\{\phi_n\}_1^\infty \subseteq \mathbf{L}$. Then $\mathbf{L}_u$ is a lattice which is closed under non-negative linear operations and non-decreasing sequential limits (i.e., $\{f_n\}_1^\infty \subseteq \mathbf{L}_u$ and $f_n \nearrow f$ implies that $f \in \mathbf{L}_u$). Moreover, I admits a unique extension to $\mathbf{L}_u$ in such a way that $I(f_n) \nearrow I(f)$ whenever f is the (pointwise) limit of a non-decreasing $\{f_n\}_1^\infty \subseteq \mathbf{L}_u$. In particular, for all $f, g \in \mathbf{L}_u$, $I(f) \leq I(g)$ if $f \leq g$ and $I(\alpha f + \beta g) = \alpha I(f) + \beta I(g)$ for all $\alpha, \beta \in [0, \infty)$.*

PROOF: The closedness properties of $\mathbf{L}_u$ are obvious. Moreover, given that an extension of I exists, it is clear that it is unique, monotone, and linear under non-negative linear operations.

Just as in the development in Section III.2 which eventually led to The Monotone Convergence Theorem, the proof (cf. Lemma III.2.6) that I extends to $\mathbf{L}_u$ is simply a matter of checking that the desired extension of I is consistent. Thus, what we must show is that when $\psi \in \mathbf{L}$ and $\{\phi_n\}_1^\infty \subseteq \mathbf{L}$ is a non-decreasing sequence with the property that $\psi \leq \lim_{n \to \infty} \phi_n$ point-wise, then $I(\psi) \leq \lim_{n \to \infty} I(\phi_n)$. To this end, set $\eta_n = \psi \wedge \phi_n$, note that $\psi - \eta_n \searrow 0$, and conclude that

$$\lim_{n \to \infty} I(\phi_n) \geq \lim_{n \to \infty} I(\eta_n) = \lim_{n \to \infty} \left(I(\psi) - I(\psi - \eta_n)\right) = I(\psi).$$

As we said before, once one knows that I is consistently defined on $\mathbf{L}_u$, the rest of the proof differs in no way from the proof of The Monotone Convergence Theorem (cf. Theorem III.3.2). ∎

The Lemma VI.1.3 gives the first step in the extension of I. What it does is provide us with a rich class functions which will be used to play the role which open sets played in our construction of LEBESGUE's measure. Thus, given any $f : E \longrightarrow \overline{\mathbb{R}}$, we define

(VI.1.4) $$\overline{I}(f) = \inf\Big\{ I(\phi) : \phi \in \mathbf{L}_u \text{ and } f \le \phi \Big\}.$$

(We use the convention that the infimum over the empty set is $+\infty$.) Clearly $\overline{I}(f)$ is the analog here of the outer measure $\Gamma \longmapsto |\Gamma|_e$ in Section II.1. At the same time as we consider $\overline{I}$, it will be convenient to have

(VI.1.5) $$\underline{I}(f) = \sup\Big\{ -I(\phi) : \phi \in \mathbf{L}_u \text{ and } -\phi \le f \Big\};$$

the analog of which in Section II.1 would have been the **interior measure**

$$\Gamma \longmapsto |\Gamma|_i \equiv \sup\Big\{ |F| : F \text{ is closed and } F \subseteq \Gamma \Big\}$$

(In keeping with our convention about the infimum, we take the supremum over the empty set to be $-\infty$.) Obviously,

(VI.1.6) $$\underline{I}(f) = -\overline{I}(-f),$$

(VI.1.7) $$\overline{I}(f) \le \overline{I}(g) \quad \text{and} \quad \underline{I}(f) \le \underline{I}(g) \quad \text{for} \quad f \le g,$$

and

(VI.1.8) $$\begin{aligned} &\overline{I}(\alpha f) = \alpha \overline{I}(f) \text{ and } \underline{I}(\alpha f) = \alpha\underline{I}(f) \quad \text{if } \alpha \in [0,\infty) \\ &\overline{I}(\alpha f) = \alpha\underline{I}(f) \text{ and } \underline{I}(\alpha f) = \alpha\overline{I}(f) \quad \text{if } \alpha \in (-\infty, 0]. \end{aligned}$$

Not quiet so obvious are the facts that

(VI.1.9) $$\underline{I}(f) \le \overline{I}(f) \text{ for all } f\text{'s} \quad \text{and} \quad \underline{I}(f) = I(f) = \overline{I}(f) \text{ for } f \in \mathbf{L}_u,$$

and that, for any $\widehat{\mathbb{R}^2}$-valued (cf. Section III.2) pair (f,g),

(VI.1.10) $$\overline{I}(f + g) \le \overline{I}(f) + \overline{I}(g) \quad \text{and} \quad \underline{I}(f + g) \ge \underline{I}(f) + \underline{I}(g)$$

so long as the right hand sides are defined. The first part of (VI.1.9) comes down to showing that if $\phi, \psi \in \mathbf{L}_u$ and $-\phi \le f \le \psi$ then $-I(\phi) \le I(\psi)$; but since $\psi + \phi \ge 0$, $I(\psi) + I(\phi) = I(\psi + \phi) \ge 0$. To see the second part of (VI.1.9), suppose that $f \in \mathbf{L}_u$. Then clearly $\overline{I}(f) \le I(f)$. At the same time, if $\{\phi_n\}_1^\infty \subseteq \mathbf{L}$ is chosen so that $\phi_n \nearrow f$, then (because $-\phi_n \in \mathbf{L} \subseteq \mathbf{L}_u$ for each $n \in \mathbb{Z}^+$)

$$\underline{I}(f) \ge \lim_{n \to \infty} -I(-\phi_n) = \lim_{n \to \infty} I(\phi_n) = I(f).$$

Turning to the proof of (VI.1.10), note that, by (VI.1.6), we need only work with $\overline{I}$. Next observe that when $\overline{I}(f) \wedge \overline{I}(g) > -\infty$, the inequality is trivial. On the other hand, if $\overline{I}(f) = -\infty$ and $\overline{I}(g) < \infty$, we can choose $\{\phi_n\}_1^\infty \subseteq \mathbf{L}_u$ so that $f \le \phi_n$ and $I(\phi_n) \le -n$, and we can find $\psi \in \mathbf{L}_u$ with $g \le \psi$ and $I(\psi) < \infty$. Hence, since $f + g \le \phi_n + \psi$, $\overline{I}(f + g) \le \lim_{n \to \infty} (-n + I(\psi)) = -\infty$.

VI.1.11 Lemma. *If $\{f_n\}_1^\infty$ is a sequence of non-negative functions on E, then*

$$(VI.1.12) \qquad \overline{I}\left(\sum_1^\infty f_n\right) \le \sum_1^\infty \overline{I}(f_n).$$

PROOF: In proving (VI.1.12), we assume, without loss in generality, that $\overline{I}(f_n) < \infty$ for every $n \in \mathbb{Z}^+$. Thus, for given $\epsilon > 0$, we can find non-negative $\phi_n \in \mathbf{L}_u$ such that $f_n \le \phi_n$ and $I(\phi_n) \le \overline{I}(f_n) + \epsilon/2^n$. Obviously, $\phi \equiv \sum_1^\infty \phi_n \ge f$; and, because the ϕ_n's are non-negative, $\phi \in \mathbf{L}_u$. Hence:

$$\overline{I}(f) \le I(\phi) = \lim_{n\to\infty} I\left(\sum_{m=1}^n \phi_m\right) = \lim_{n\to\infty} \sum_{m=1}^n I(\phi_m) \le \sum_{m=1}^\infty \overline{I}(f_m) + \epsilon. \quad \blacksquare$$

With the preceding preparations in hand, we are now ready to complete our extension program. Namely, define $\overline{L^1(I)}$ to be the set of $f : E \longrightarrow \overline{\mathbb{R}}$ with the property that $\underline{I}(f) = \overline{I}(f) \in \mathbb{R}$; and, for $f \in \overline{L^1(I)}$, define $\tilde{I}(f) = \overline{I}(f)$. Finally, let $L^1(I)$ be the set of $\mathbb{R}$-valued elements of $\overline{L^1(I)}$.

VI.1.13 Theorem. (DANIELL) *Let $(E, \mathbf{L}, I)$ be an integration theory. Then $(E, L^1(I), \tilde{I})$ is again an integration theory, $\mathbf{L} \subseteq L^1(I)$, and $\tilde{I}$ agrees with I on $\mathbf{L}$. Moreover, if $\{f_n\}_1^\infty \subseteq \overline{L^1(I)}$ and $f_n \nearrow f$, then*

$$(VI.1.14) \qquad \tilde{I}(f_n) \nearrow \underline{I}(f) = \overline{I}(f);$$

and so, $f \in \overline{L^1(I)}$ if and only if $\overline{I}(f) \in \mathbb{R}$. Finally, if $\mathbf{L}_{u\ell}$ is the set of $f : E \longrightarrow \overline{\mathbb{R}}$ which can be written as the limit of a non-increasing sequence from $\mathbf{L}_u$, then an element f of $\mathbf{L}_{u\ell}$ is an element of $\overline{L^1(I)}$ if and only if $\overline{I}(f) \in \mathbb{R}$. In fact, for any $f : E \longrightarrow \overline{\mathbb{R}}$, $f \in \overline{L^1(I)}$ if and only if there exist $\phi, \psi \in \mathbf{L}_{u\ell} \cap \overline{L^1(I)}$ with the properties that $-\phi \le f \le \psi$ and $-\tilde{I}(\phi) = \tilde{I}(\psi)$.

PROOF: We must first check that $L^1(I)$ is a vector lattice and that $\tilde{I}$ is linear on $L^1(I)$ (it is obviously non-negative). But (VI.1.8) and (VI.1.10) clearly show that $L^1(I)$ is a vector space and that $\tilde{I}$ is linear there. Thus, all that we have to do is show that $f \vee 0 \in L^1(I)$ whenever $f \in L^1(I)$. To this end, let $\epsilon > 0$ be given and choose $\phi, \psi \in \mathbf{L}_u \cap \overline{L^1(I)}$ so that $-\phi \le f \le \psi$ and

$$\left(I(\psi) - \tilde{I}(f)\right) \vee \left(\tilde{I}(f) + I(\phi)\right) < \frac{\epsilon}{2}.$$

Since both $\phi \wedge 0$ and $\psi \vee 0$ are again elements of $\mathbf{L}_u$ and $-(\phi \wedge 0) \le f \vee 0 \le \psi \vee 0$, we see that

$$-I(\phi \wedge 0) \le \underline{I}(f \vee 0) \le \overline{I}(f \vee 0) \le I(\psi \vee 0).$$

In addition, since $\phi + \psi \geq 0$, $\phi \wedge 0 + \psi \vee 0 \leq \phi + \psi$, and therefore $I(\phi \wedge 0) + I(\psi \vee 0) < \epsilon$. Hence

$$-I(\phi \wedge 0) \leq \underline{I}(f \vee 0) \leq \overline{I}(f \vee 0) \leq I(\psi \vee 0) \leq -I(\phi \wedge 0) + \epsilon;$$

and so it remains only to check that $I(\phi \wedge 0) \in \mathbb{R}$. But, since $\phi \wedge 0 \in \mathbf{L}_u$, $I(\phi \wedge 0) > -\infty$; and because $\phi \wedge 0 \leq \phi \in \overline{L^1(I)}$, $I(\phi \wedge 0) < \infty$.

We next prove (VI.1.14). Since $\tilde{I}(f_n) \leq \underline{I}(f) \leq \overline{I}(f)$ for every $n \in \mathbb{Z}^+$, this comes down to checking that $\overline{I}(f) \leq \lim_{n \to \infty} \tilde{I}(f_n)$; and, after replacing f_n by $f_n \wedge n$ if necessary, we see that it suffices to do this in the case when the f_n's take values in $[-\infty, \infty)$. But in that case, set $h_1 = f_1$ and $h_{n+1} = f_{n+1} - f_n \geq 0$, $n \in \mathbb{Z}^+$; and note that

$$f_n = \sum_{m=1}^{n} h_m \nearrow \sum_{m=1}^{\infty} h_m = f.$$

Hence, by (VI.1.10) and Lemma VI.1.11,

$$\lim_{n \to \infty} \tilde{I}(f_n) = \lim_{n \to \infty} \sum_{m=1}^{n} \tilde{I}(h_m) = \sum_{m=1}^{\infty} \overline{I}(h_m) \geq \overline{I}(f).$$

In particular, it is now clear that $f \in \overline{L^1(I)}$ if and only if $\overline{I}(f) < \infty$.

Given the preceding, it is clear that if $\{g_n\}_1^{\infty} \subseteq L^1(I)$ and $g_n \searrow 0$, then, with $f_n = g_1 - g_n$,

$$\tilde{I}(g_n) = \tilde{I}(g_1) - \tilde{I}(f_n) \searrow \tilde{I}(g_1) - \tilde{I}(g_1) = 0.$$

In other words, $\tilde{I}$ is an integral on $L^1(I)$. Furthermore, the facts that $\mathbf{L} \subseteq L^1(I)$ and that $\tilde{I}(f) = I(f)$ for $f \in \mathbf{L}$ are immediate consequences of the second part of (VI.1.9).

To see that $f \in \overline{L^1(I)}$ when $f \in \mathbf{L}_{u\ell}$ and $\overline{I}(f) \in \mathbb{R}$, choose $\{\phi\} \cup \{\phi_n\}_1^{\infty} \subseteq \mathbf{L}_u$ so that $f \leq \phi$, $I(\phi) \leq \overline{I}(f) + 1$, and $\phi_n \searrow f$. Next, set $\psi_n = \phi \wedge \phi_n$ and observe both that $\{\psi_n\}_1^{\infty} \subseteq \overline{L^1(I)}$ and that $\psi_n \searrow f$. Hence $\{-\psi_n\}_1^{\infty} \subseteq \overline{L^1(I)}$, $-\psi_n \nearrow -f$, and therefore (by the preceding) $f = -(-f) \in \overline{L^1(I)}$.

Finally, turning to the final assertion, let $f : E \longrightarrow \overline{\mathbb{R}}$. If there exist $\phi, \psi \in \mathbf{L}_{u\ell} \cap \overline{L^1(I)}$ satisfying $-\phi \leq f \leq \psi$ and $-\tilde{I}(\phi) = \tilde{I}(\psi)$, then it is clear that $f \in \overline{L^1(I)}$. On the other hand, if $f \in \overline{L^1(I)}$, then we can choose $\{\psi_n\}_1^{\infty} \subseteq \mathbf{L}_u \cap \overline{L^1(I)}$ so that $f \leq \psi_n$, $n \in \mathbb{Z}^+$, and $I(\psi_n) \searrow \tilde{I}(f)$. Clearly, after replacing ψ_n by $\psi_1 \wedge \cdots \wedge \psi_n$ if necessary, we may and will assume that the $\{\psi_n\}_1^{\infty}$ is non-increasing. Thus, if $\psi = \lim_{n \to \infty} \psi_n$, then $\psi \in \mathbf{L}_{u\ell} \cap \overline{L^1(I)}$, $f \leq \psi$, and $\tilde{I}(\psi) = \tilde{I}(f)$. By replacing f with $-f$, we can also find a $\phi \in \mathbf{L}_{u\ell} \cap \overline{L^1(I)}$ which satisfies $-\phi \leq f$ and $-\tilde{I}(\phi) = \tilde{I}(f)$. ∎

We are now ready to return to the problem, raised in the discussion following Examples VI.1.1, of identifying of the measure underlying a given integration theory $(E, \mathbf{L}, I)$. For this purpose, we introduce the notation $\sigma(\mathbf{L})$ to denote the smallest σ-algebra over E with respect to which all of the functions in the vector lattice $\mathbf{L}$ are measurable. Obviously, $\sigma(\mathbf{L})$ is generated by the sets $\{f > a\}$ as f runs over $\mathbf{L}$ and a runs over $\mathbb{R}$.

VI.1.15 Theorem. (STONE) *Let $(E, \mathbf{L}, I)$ be an integration theory. Then every element of $\mathbf{L}_{u\ell}$ is $\sigma(\mathbf{L})$-measurable. In addition, if $\mathbf{1} \in \mathbf{L}$, then*

$$(VI.1.16) \qquad \sigma\big(L^1(I)\big) = \Big\{ \Gamma \subseteq E : \chi_\Gamma \in L^1(I) \Big\},$$

the mapping

$$(VI.1.17) \qquad \Gamma \in \sigma\big(L^1(I)\big) \longmapsto \mu_I(\Gamma) \equiv \tilde{I}(\chi_\Gamma) \in [0, \infty)$$

is a finite measure on $\big(E, \sigma(L^1(I))\big)$, $\sigma\big(L^1(I)\big)$ is the completion of $\sigma(\mathbf{L})$ with respect to μ_I, $L^1(\mu_I) = L^1(I)$, and

$$(VI.1.18) \qquad \tilde{I}(f) = \int_E f \, d\mu_I, \qquad f \in L^1(I).$$

Finally, if $\mathbf{1} \in \mathbf{L}$ and $(E, \mathcal{B}, \nu)$ is any finite measure space with the properties that $\mathbf{L} \subseteq L^1(\nu)$ and $I(f) = \int_E f \, d\nu$ for every $f \in \mathbf{L}$, then $\sigma(\mathbf{L}) \subseteq \mathcal{B}$ and ν coincides with μ_I on $\sigma(\mathbf{L})$.

PROOF: Since the initial assertion is trivial, we will assume throughout that $\mathbf{1} \in \mathbf{L}$.

Let $\mathcal{H}$ denote the collection described on the right hand side of (VI.1.16). Using Theorem VI.1.13, one can easily show that $\mathcal{H}$ is a σ-algebra over E and that

$$\Gamma \in \mathcal{H} \longmapsto \mu_I(\Gamma) \equiv \tilde{I}(\chi_\Gamma) \in [0, \infty)$$

defines a finite measure on $(E, \mathcal{H})$. Our first goal is to prove that

$$(VI.1.19) \qquad L^1(I) = L^1(\mu_I) \quad \text{and} \quad \tilde{I}(f) = \int_E f \, d\mu_I \text{ for all } f \in L^1(I).$$

To this end, for given $f : E \longrightarrow \mathbb{R}$ and $a \in \mathbb{R}$, consider the functions

$$(VI.1.20) \qquad g_n \equiv \Big[n(f - a)^+ \Big] \wedge \mathbf{1}, \qquad n \in \mathbb{Z}^+.$$

If $f \in L^1(I)$, then each g_n is also an element of $L^1(I)$, $g_n \nearrow \chi_{\{f>a\}}$ as $n \longrightarrow \infty$, and therefore $\chi_{\{f>a\}} \in L^1(I)$. Thus we see that every $f \in L^1(I)$ is $\mathcal{H}$-measurable. Next, for given $f : E \longrightarrow [0,\infty)$ define

$$f_n = \sum_{k=0}^{4^n} \frac{k}{2^n} \chi_{\{k \leq 2^n f < k+1\}} \quad \text{for} \quad n \in \mathbb{Z}^+.$$

If $f \in L^1(I) \cup L^1(\mu_I)$, then $f_n \in L^1(I) \cap L^1(\mu_I)$, $f_n \nearrow f$, and so $f \in L^1(I) \cap L^1(\mu_I)$ and $\tilde{I}(f) = \int_E f \, d\mu_I$. Hence, we have now proved (VI.1.19).

Our next goal is to show that

(VI.1.21) $$\overline{\sigma(\mathbf{L})}^{\mu_I} = \mathcal{H} = \sigma(L^1(I)).$$

Since $\mathbf{L} \subseteq L^1(I)$ and every element of $L^1(I)$ is $\mathcal{H}$-measurable, what we know so far is that

$$\sigma(\mathbf{L}) \subseteq \sigma(L^1(I)) \subseteq \mathcal{H}.$$

Thus, to prove (VI.1.21) all that we need to do is show that

$$\overline{\mathcal{H}}^{\mu_I} \subseteq \overline{\sigma(\mathbf{L})}^{\mu_I} \cap \sigma(L^1(I)).$$

But if $\Gamma \in \overline{\mathcal{H}}^{\mu_I}$, then there exist $\phi, \psi \in \mathbf{L}_{u\ell} \cap L^1(I)$ with the properties that $-\phi \leq \chi_\Gamma \leq \psi$ and $\tilde{I}(-\phi) = \mu(\Gamma) = \tilde{I}(\psi)$. Thus Γ is certainly an element of $\sigma(L^1(I))$. Moreover, if $A = \{\phi < 0\}$ and $B = \{\psi \geq 1\}$, then A and B are both elements of $\sigma(\mathbf{L})$ and $-\phi \leq \chi_A \leq \chi_\Gamma \leq \chi_B \leq \psi$. In particular, $A \subseteq \Gamma \subseteq B$ and $\mu_I(A) = \mu_I(B)$, and therefore Γ is also an element of $\overline{\sigma(\mathbf{L})}^{\mu_I}$.

We have now completed the proof of everything except the concluding assertion of uniqueness. But if $\mathbf{L} \subseteq L^1(\nu)$, then obviously $\sigma(\mathbf{L}) \subseteq \mathcal{B}$. Moreover, if $I(f) = \int f \, d\nu$ for all $f \in \mathbf{L}$, then by again considering the functions g_n, $n \in \mathbb{Z}^+$, in (VI.1.20) we see that

$$\mu_I(f > a) = \nu(f > a) \quad \text{for } f \in \mathbf{L} \text{ and } a \in \mathbb{R}.$$

Thus, by Exercise III.1.8, ν and μ_I are equal on $\sigma(\mathbf{L})$. ∎

With the preceding result, we are now ready to handle the situations described in ii) and iii) of Examples IV.1.1.

VI.1.22 Corollary. (CARATHÉODORY EXTENSION) *Let $\mathcal{A}$ be an algebra of subsets of E and suppose that $\mu : \mathcal{A} \longrightarrow [0,\infty)$ is a finitely additive function (cf. ii) in Examples VI.1.1) with the property that (VI.1.2) holds. Then there is a unique finite measure $\tilde{\mu}$ on $\big(E, \sigma(\mathcal{A})\big)$ with the property that $\tilde{\mu}$ coincides with μ on $\mathcal{A}$.*

PROOF: Define $\mathbf{L}(\mathcal{A})$ and I on $\mathbf{L}(\mathcal{A})$ as in ii) of Examples VI.1.1. It is then an easy matter to see that $\sigma(\mathcal{A}) = \sigma\big(\mathbf{L}(\mathcal{A})\big)$. In addition, as was shown in ii) of Examples VI.1.1, I is an integral on $\mathbf{L}(\mathcal{A})$. Hence the desired existence and uniqueness statements follow immediately from Theorem VI.1.15. ∎

Before we can complete iii) in Examples VI.1.1, we must first prove the lemma alluded to there.

VI.1.23 Lemma. (DINI'S LEMMA) *Let $\big\{f_n\big\}_1^\infty$ be a non-increasing sequence of non-negative, continuous functions on the topological space E. If $f_n \searrow 0$, then $f_n \longrightarrow 0$ uniformly on each compact subset $K \subseteq E$ (i.e., $\|f_n\|_{u,K} \longrightarrow 0$.)*

PROOF: Without loss in generality, we assume that E itself is compact.

Let $\epsilon > 0$ be given. By assumption, we can find for each $x \in E$ an $n(x) \in \mathbb{Z}^+$ and an open neighborhood $U(x)$ of x such that $f_{n(x)}(y) \leq \epsilon$ for all $y \in U(x)$. Moreover, by the HEINE-BOREL Theorem, we can choose a finite set $\{x_1, \ldots, x_L\} \subseteq E$ so that $E = \bigcup_{\ell=1}^L U(x_\ell)$. Thus, if $N(\epsilon) = n(x_1) \vee \cdots \vee n(x_L)$, then $f_n \leq \epsilon$ so long as $n \geq N(\epsilon)$. ∎

Given a topological space E, let $C_b(E;\mathbb{R})$ denote the space of bounded continuous functions on E and turn $C_b(E;\mathbb{R})$ into a metric space by defining $\|g - f\|_{u,E}$ to be the distance between f and g. We will say that $\Lambda : C_b(E;\mathbb{R}) \longrightarrow \mathbb{R}$ is a **non-negative linear functional** if Λ is linear and $\Lambda(f) \geq 0$ for all $f \in C_b\big(E;[0,\infty)\big)$. Furthermore, if Λ is a non-negative linear functional on $C_b(E;\mathbb{R})$, we will say that Λ is **tight** if it has the property that for every $\delta > 0$ there is a compact $K_\delta \subseteq E$ and an $A_\delta \in (0,\infty)$ for which

$$|\Lambda(f)| \leq A_\delta \|f\|_{u,K_\delta} + \delta \|f\|_{u,E} \quad \text{for all } f \in C_b(E;\mathbb{R}).$$

Notice that when E is itself compact, then every non-negative linear functional on $C_b(E;\mathbb{R})$ is tight.

VI.1.24 Theorem. (RIESZ REPRESENTATION) *Let E be a topological space, set $\mathcal{B} = \sigma\big(C_b(E;\mathbb{R})\big)$, and suppose that $\Lambda : C_b(E;\mathbb{R}) \longrightarrow \mathbb{R}$ is a non-negative* *linear functional which is tight. Then there is a unique finite measure μ on $(E, \mathcal{B})$ with the property that $\Lambda(f) = \int_E f \, d\mu$, $f \in C_b(E; \mathbb{R})$.*

PROOF: Clearly all that we need to do is show that $\Lambda(f_n) \searrow 0$ whenever $\{f_n\}_1^\infty \subseteq C_b(E; \mathbb{R})$ is a non-increasing sequence which tends to 0. To this end, let $\epsilon > 0$ be given, set $\delta = \frac{\epsilon}{1 + 2\|f_1\|_{u,E}}$, and use DINI's Lemma to choose an $N(\delta) \in \mathbb{Z}^+$ so that $\|f_n\|_{u,K_\delta} \leq \epsilon/A_\delta$ for all $n \geq N(\delta)$, where K_δ and A_δ are the quantities appearing in the tightness condition for Λ. Then, for $n \geq N(\delta)$, $|\Lambda(f_n)| \leq 2\epsilon$. ∎

VI.1.25 Exercise.

Assume that E is a metric space. Show that $\sigma(C_b(E; \mathbb{R}))$ coincides with the BOREL field $\mathcal{B}_E$. Next, assume in addition that E is locally compact (i.e. every point $x \in E$ has a neighborhood whose closure is compact) and show that $\mathcal{B}_E = \sigma(C_c(E; \mathbb{R}))$ (the space of $f \in C(E; \mathbb{R})$ which vanish off of some compact subset of E).

VI.1.26 Exercise.

Let $\mathbf{L}$ be the space of functions $f : \mathbb{R} \longrightarrow \mathbb{R}$ with the property that $f = c + \phi$ for some $c \in \mathbb{R}$ and $\phi \in C_c(\mathbb{R}; \mathbb{R})$. Show that $\mathbf{L}$ is a vector lattice. Next, let $\psi : \mathbb{R} \longrightarrow \mathbb{R}$ be a bounded, right-continuous, non-decreasing function and set

$$\psi(-\infty) \equiv \lim_{t \searrow -\infty} \psi(t) > -\infty \quad \text{and} \quad \psi(\infty) \equiv \lim_{t \nearrow \infty} \psi(t) < \infty.$$

Show that the functional $I : \mathbf{L} \longrightarrow \mathbb{R}$ given by

$$I(f) = c(\psi(\infty) - \psi(-\infty)) + (R) \int_{[a,b]} \phi(t) \, d\psi(t),$$

when $-\infty < a < b < \infty$ and $f = c + \phi$, with $\phi \in C_c(\mathbb{R}; \mathbb{R})$ vanishing off of $[a, b]$, is well-defined, linear, and non-negative. Finally, show that I is an integral on $\mathbf{L}$, and conclude that there is a unique finite measure μ_ψ on $(\mathbb{R}, \mathcal{B}_\mathbb{R})$ with the property that

$$\mu_\psi((a, b]) = \psi(b) - \psi(a) \quad \text{for all} \quad -\infty < a \leq b < \infty.$$

Notice that the mapping $\psi \longmapsto \mu_\psi$ described here inverts the map discussed in Theorem IV.1.12.

Hint: Show that if $f = c + \phi$, where $\phi \in C_c(\mathbb{R}; \mathbb{R})$ vanishes outside of $[a, b]$, then:

$$I(f) = c\big(\psi(\infty) - \psi(b)\big) + c\big(\psi(a) - \psi(-\infty)\big) + (R) \int_{[a,b]} f(t) \, d\psi(t).$$

VI.2. Hilbert Space and the Radon–Nikodym Theorem.

In Exercise V.1.7 we saw evidence that, among the L^p-spaces, the space L^2 is the most closely related to familiar EUCLIDean geometry. In the present section, we will expand on the this observation and give an application of it.

Throughout $(E, \mathcal{B}, \mu)$ will be a measure space. As we saw in part **iii)** of Theorem V.2.2, the space $L^2(\mu)$ is a vector space which becomes a complete metric space when we use $\|f - g\|_{L^2(\mu)}$ to measure the distance between f and g. In addition, if we define

(VI.2.1) $(f, g) \in \big(L^2(\mu)\big)^2 \longmapsto (f, g)_{L^2(\mu)} \equiv \int_E f \cdot g \, d\mu \in \mathbb{R},$

then $(f, g)_{L^2(\mu)}$ is bilinear (i.e., it is linear in each of its variables) and, for $f \in L^2(\mu)$,

(VI.2.2)
$$\|f\|_{L^2(\mu)} = (f, f)_{L^2(\mu)}^{1/2}$$
$$= \sup\Big\{ (f, g)_{L^2(\mu)} : g \in L^2(\mu) \text{ with } \|g\|_{L^2(\mu)} \leq 1 \Big\}.$$

Note that $(f, g)_{L^2(\mu)}$ plays the same role for $L^2(\mu)$ that the EUCLIDean inner product plays in $\mathbb{R}^N$. Thus, we say that $f \in L^2(\mu)$ is **orthogonal** or **perpendicular** to $S \subseteq L^2(\mu)$ and write $f \perp S$ if $(f, g)_{L^2(\mu)} = 0$ for every $g \in S$.

A topological vector spaces with this kind of structure is known as a **Hilbert space**. A distinguishing feature of HILBERT spaces is the property proved in the following lemma. Indeed, the reader might want to observe that, although we restrict our attention to L^2-spaces, the result proved depends only on completeness and the existence of a bilinear map for which (VI.2.2) holds; hence, it is a property possessed by all HILBERT spaces.

VI.2.3 Lemma. *Let $F \neq \emptyset$ be a closed linear subspace of $L^2(\mu)$. Then, for each $g \in L^2(\mu)$, there is a unique $f \in F$ for which*

(VI.2.4) $$\|g - f\|_{L^2(\mu)} = \inf\Big\{\|g - \phi\|_{L^2(\mu)} : \phi \in F\Big\}.$$

In fact, the unique $f \in F$ which satisfies (VI.2.4) is the unique $f \in F$ such that $(g - f) \perp F$.

PROOF: We first check that $f \in F$ satisfies (VI.2.4) if and only if $(g - f) \perp F$. To this end, suppose that $f \in F$ satisfies (VI.2.4), and observe that, for any $\psi \in F$, the function

$$t \in \mathbb{R} \longmapsto \big\|g - f - t\psi\big\|^2_{L^2(\mu)} = \|g - f\|^2_{L^2(\mu)} - 2t(g - f, \psi)_{L^2(\mu)} + t^2\|\psi\|^2_{L^2(\mu)}$$

has a minimum at $t = 0$. Hence, by the first derivative test, we see that $(g - f, \psi)_{L^2(\mu)} = 0$ for every $\psi \in F$. Conversely, if $f \in F$ and $(g - f) \perp F$, then, for any $\psi \in F$

$$\|g - \psi\|^2_{L^2(\mu)} = \|g - f\|^2_{L^2(\mu)} + 2(g - f, f - \psi)_{L^2(\mu)} + \|f - \psi\|^2_{L^2(\mu)}$$
$$= \|g - f\|^2_{L^2(\mu)} + \|f - \psi\|^2_{L^2(\mu)} \geq \|g - f\|^2_{L^2(\mu)}.$$

Thus, we have now proved the equivalence of the two characterizations of f; and, as consequence of the second characterization, uniqueness is easy. Indeed, if $f_1, f_2 \in F$ and $(g - f_i) \perp F$, $i \in \{1, 2\}$, then $(f_2 - f_1) \perp F$ and therefore $\big\|f_2 - f_1\big\|_{L^2(\mu)} = 0$.

In view of the preceding, it remains only to prove that there is an $f \in F$ for which (VI.2.4) holds. To this end, choose $\{f_n\}_1^\infty \subseteq F$ so that

$$\|g - f_n\|_{L^2(\mu)} \longrightarrow \alpha \equiv \inf\Big\{\|g - \phi\|_{L^2(\mu)} : \phi \in F\Big\}.$$

Since $L^2(\mu)$ and therefore F are complete, all that we have to do is show that $\{f_n\}_1^\infty$ is CAUCHY convergent. In order to do this, note that for any $\phi, \psi \in L^2(\mu)$,

(VI.2.5) $$\|\phi + \psi\|^2_{L^2(\mu)} + \|\phi - \psi\|^2_{L^2(\mu)} = 2\|\phi\|^2_{L^2(\mu)} + 2\|\psi\|^2_{L^2(\mu)}.$$

Taking $\phi = g - f_n$ and $\psi = g - f_m$ in (VI.2.5), we obtain

$$\|f_n - f_m\|^2_{L^2(\mu)} = 2\|g - f_n\|^2_{L^2(\mu)} + \|g - f_m\|^2_{L^2(\mu)} - 4\left\|g - \frac{f_n + f_m}{2}\right\|^2_{L^2(\mu)}$$

$$\leq 2\Big(\|g - f_n\|^2_{L^2(\mu)} - \alpha^2\Big) + 2\Big(\|g - f_m\|^2_{L^2(\mu)} - \alpha^2\Big),$$

where we have used the fact that $\frac{f_n + f_m}{2} \in F$ in order to get the last inequality. Now let $\epsilon > 0$ be given, choose $N \in \mathbb{Z}^+$ so that $\|g - f_n\|^2_{L^2(\mu)} < \alpha^2 + \epsilon^2/4$ for $n \geq N$, and conclude that $\|f_n - f_m\|_{L^2(\mu)} < \epsilon$ for $m, n \geq N$. ∎

The result obtained in Lemma VI.1.3 is a basic existence assertion from which a great many other existence results follow. Perhaps the single most important such result is the following sharpening of the (V.2.8).

VI.2.6 Theorem. (F. RIESZ) *Let* $\Lambda : L^2(\mu) \longrightarrow \mathbb{R}$ *be a linear mapping. Then*

$$(VI.2.7) \qquad \sup\Big\{\Lambda(\phi) : \phi \in L^2(\mu) \text{ and } \|\phi\|_{L^2(\mu)} \le 1\Big\} < \infty$$

if and only if there is an $h \in L^2(\mu)$ *for which*

$$(VI.2.8) \qquad \Lambda(\phi) = \big(h, \phi\big)_{L^2(\mu)}, \qquad \phi \in L^2(\mu);$$

in which case h *is uniquely determined by* (VI.2.8) *and* $\|h\|_{L^2(\mu)}$ *is the supremum in* (VI.2.7).

PROOF: Everything except the existence of f when (VI.2.7) holds is clear. To prove this existence result, assume that (VI.2.7) holds. If $\Lambda(\phi) = 0$, $\phi \in L^2(\mu)$, then we can simply take $h = 0$. Thus, we assume that $\Lambda(\phi) \ne 0$ for some $\phi \in L^2(\mu)$, and we define $F = \{\phi \in L^2(\mu) : \Lambda(\phi) = 0\}$. Obviously F is a linear subspace and, by (VI.2.7), F is closed. Moreover, by assumption, there is an $g \notin F$. Now choose $f \in F$ so that $k \equiv (g - f) \perp F$ and note that $\Lambda(k) \ne 0$. In addition, observe that, for any $\phi \in L^2(\mu)$,

$$\Lambda\left(\phi - \frac{\Lambda(\phi)}{\Lambda(k)}k\right) = 0$$

and therefore that $\phi - \frac{\Lambda(\phi)}{\Lambda(k)}k \in F$. Hence,

$$(k, \phi)_{L^2(\mu)} - \frac{\Lambda(\phi)}{\Lambda(k)}\|k\|^2_{L^2(\mu)} = \left(k, \phi - \frac{\Lambda(\phi)}{\Lambda(k)}k\right) = 0, \qquad \phi \in L^2(\mu);$$

and so we can take

$$h = \frac{\Lambda(k)}{\|k\|^2_{L^2(\mu)}}k. \quad \blacksquare$$

Following J. VON NEUMANN, we will now use Theorem VI.2.6 to derive an important property about the relationship between measures. Namely, given two measures μ and ν on the same measurable space $(E, \mathcal{B})$:

a) we say ν **dominates** μ and write $\mu \le \nu$ if $\mu(\Gamma) \le \nu(\Gamma)$, $\Gamma \in \mathcal{B}$;

b) we say μ is **absolutely continuous with respect to** ν and write $\mu \ll \nu$ if $\mu(\Gamma) = 0$ for all $\Gamma \in \mathcal{B}$ with $\nu(\Gamma) = 0$;

c) we say μ and ν are **singular** and write $\mu \perp \nu$ if there is a $\Sigma \in \mathcal{B}$ with the property that $\mu(\Sigma) = \nu(\Sigma^\complement) = 0$.

Obviously, both domination and absolute continuity express a relationship between μ and ν; the former being a much stronger statement than the latter. In contrast, singularity is a statement that the measures have nothing to do with one another and, in fact, *live on different portions of E.*

The result alluded to above comes in two parts and applies to μ's which are finite and ν's which are σ-finite. The first part, which is called LEBESGUE's Decomposition Theorem, says that μ can be written (in a unique way) as the sum of a measure μ_a which is absolutely continuous with respect to ν and a measure μ_σ which is singular to ν. The second part, which is known as the RADON–NIKODYM Theorem, tells us that there is a unique non-negative $f \in L^1(\nu)$ with the property that

$$(\text{VI.2.9}) \qquad \mu_a(\Gamma) = \int_\Gamma f\, d\nu, \qquad \Gamma \in \mathcal{B}.$$

In particular, if μ itself is absolutely continuous with respect to ν, then $\mu = \mu_a$ and so (VI.2.9) holds with μ in place of μ_a.

The key to VON NEUMANN's proof of these results is the observation that everything can be reduced to consideration μ's which are dominated by ν; in which case the RADON–NIKODYM Theorem becomes a simple application of Theorem VI.2.6.

VI.2.10 Lemma. *Suppose that $(E, \mathcal{B}, \nu)$ is a σ-finite measure space and that μ is a finite measure on $(E, \mathcal{B})$ which is dominated by ν. Then there is a unique $[0, 1]$-valued $h \in L^1(\nu)$ with the property that*

$$(\text{VI.2.11}) \qquad \int_E \phi\, d\mu = \int_E \phi \cdot h\, d\nu$$

for every $\mathcal{B}$-measurable $\phi : E \longrightarrow [0, \infty]$.

PROOF: Since we can write E as the union of countably many mutually disjoint $\mathcal{B}$-measurable sets of finite ν-measure, we assume, without loss in generality, that ν is finite on $(E, \mathcal{B})$. But, in that case, $L^1(\mu) \subseteq L^2(\nu)$ and the linear mapping

$$\phi \in L^2(\nu) \longrightarrow \Lambda(\phi) \equiv \int_E \phi\, d\mu \in \mathbb{R}$$

satisfies $\left|\Lambda(\phi)\right| \le \nu(E)^{1/2}\|\phi\|_{L^2(\nu)}$. Hence, by Theorem VI.2.6, there is an $h \in L^2(\nu)$ such that (VI.2.11) holds for every $\phi \in L^2(\nu)$ and, therefore, for every bounded $\mathcal{B}$-measurable function. We now want to show h (which is determined only up to a set of ν-measure 0) can be chosen to take its values in $[0,1]$. To this end, set $A_n = \{h \le -1/n\}$ and $B_n = \{h \ge 1 + 1/n\}$ for $n \in \mathbb{Z}^+$. Then, by (VI.2.11),

$$-\frac{1}{n}\nu(A_n) \ge \mu(A_n) \ge 0 \quad \text{and} \quad \left(1 + \frac{1}{n}\right)\nu(B_n) \le \mu(B_n) \le \nu(B_n);$$

from which we conclude that $\nu(A_n) = \nu(B_n) = 0$, $n \in \mathbb{Z}^+$, and therefore that $\nu(h < 0) = \nu(h > 1) = 0$. In other words, we may assume that h takes its values in $[0,1]$; and, clearly, once we know this, (VI.2.11) for all non-negative, $\mathcal{B}$-measurable ϕ's is an easy consequence of the Monotone Convergence Theorem.

Finally, the uniqueness assertion is trivial, since (VI.1.11) determines the ν-integral of h over every $\Gamma \in \mathcal{B}$. ∎

VI.2.12 Theorem. (THE LEBESGUE DECOMPOSITION AND RADON–NIKO-DYM THEOREMS) *Suppose that $(E, \mathcal{B}, \nu)$ is a σ-finite measure space and let μ be a finite measure on $(E, \mathcal{B})$. Then there is a unique measure $\mu_a \le \mu$ on $(E, \mathcal{B})$ with the properties that $\mu_a << \nu$ and $\mu_\sigma \equiv (\mu - \mu_a) \perp \nu$. In addition, there is a unique non-negative $f \in L^1(\nu)$ for which (VI.2.9) holds. In particular, $\mu << \nu$ if and only $\mu(\Gamma) = \int_\Gamma f\,d\nu$, $\Gamma \in \mathcal{B}$, for some non-negative $f \in L^1(\mu)$.*

PROOF: We first note that if $\mu(\Gamma) = \int_\Gamma f\,d\nu$, $\Gamma \in \mathcal{B}$, for some $f \in L^1(\nu)$, then (cf. Exercise III.3.15) $\mu << \nu$ and f is necessarily unique and non-negative as an element of $L^1(\nu)$. Next, we prove that there is at most one choice of μ_a. To this end, suppose that $\mu = \mu_a + \mu_\sigma = \mu'_a + \mu'_\sigma$, where μ_a and μ'_a are both absolutely continuous with respect to ν and both μ_σ and μ'_σ are singular to ν. Choose $\Sigma, \Sigma' \in \mathcal{B}$ so that

$$\nu(\Sigma^\complement) = \nu\left((\Sigma')^\complement\right) = 0 \quad \text{and} \quad \mu_\sigma(\Sigma) = \mu'_\sigma(\Sigma') = 0,$$

and set $A = \Sigma \cap \Sigma'$. Then $\nu(A^\complement) = \mu_\sigma(A) = \mu'_\sigma(A) = 0$; and therefore, for any $\Gamma \in \mathcal{B}$,

$$\mu_a(\Gamma) = \mu_a(A \cap \Gamma) = \mu(A \cap \Gamma) = \mu'_a(A \cap \Gamma) = \mu'_a(\Gamma).$$

Hence, $\mu_a = \mu'_a$.

To prove the existence statements, we first use Lemma VI.2.10, applied to μ and $\mu + \nu$, to find a $\mathcal{B}$-measurable $h : E \longrightarrow [0,1]$ with the property that

$$\int_E \phi\,d\mu = \int_E \phi \cdot h\,d\mu + \int_E \phi \cdot h\,d\nu$$

for all non-negative, $\mathcal{B}$-measurable ϕ's. It is then clear that

(VI.2.13)
$$\int_E \phi \cdot (1 - h)\, d\mu = \int_E \phi \cdot h\, d\nu,$$

first for all $\phi \in L^1(\mu)$ and then for all non-negative, $\mathcal{B}$-measurable ϕ's. Now set $\Sigma = \{h < 1\}$ and $\mu_a(\Gamma) = \mu(\Sigma \cap \Gamma)$, $\Gamma \in \mathcal{B}$. Since

$$\nu(\Sigma^\complement) = \nu(h = 1) = \int_{\{h=1\}} h\, d\nu = \int_{\{h=1\}} (1 - h)\, d\mu = 0,$$

it is clear that $\mu - \mu_a$ is singular to ν. At the same time, if

$$g_n \equiv \frac{1}{1 + \frac{1}{n} - h}, \; n \in \mathbb{Z}^+, \quad \text{and} \quad f \equiv \frac{h \cdot \chi_\Sigma}{1 - h},$$

then $0 \leq g_n \nearrow \frac{1}{1-h}$ point-wise on Σ and, by The Monotone Convergence Theorem:

$$\mu_a(\Gamma) = \lim_{n \to \infty} \int_{\Sigma \cap \Gamma} g_n \cdot (1 - h)\, d\mu = \lim_{n \to \infty} \int_{\Sigma \cap \Gamma} g_n \cdot h\, d\nu = \int_\Gamma f\, d\nu. \; \blacksquare$$

Given a finite measure μ and a σ-finite measure ν, the corresponding measures μ_a and μ_σ are called the **absolutely continuous** and **singular parts of μ with respect to ν**. Also, if μ is absolutely continuous with respect to ν, then the corresponding non-negative $f \in L^1(\nu)$ is called the **Radon–Nikodym derivative of μ with respect to ν** and is often denoted by $\frac{d\mu}{d\nu}$. The choice of this notation is explained by part **ii)** of the exercise which follows.

VI.2.14 Exercise.

Suppose that $\mathcal{C}$ is a countable partition of the non-empty set E, and use $\mathcal{B}$ to denote $\sigma(\mathcal{C})$.

i) Show that $f : E \longrightarrow \overline{\mathbb{R}}$ is $\mathcal{B}$-measurable if and only if f is constant on each $\Gamma \in \mathcal{C}$. Also, show that the measure ν is σ-finite on $(E, \mathcal{B})$ if and only if $\nu(\Gamma) < \infty$ for every $\Gamma \in \mathcal{C}$. Finally, if μ is a second measure on $(E, \mathcal{B})$, show that $\mu << \nu$ if and only if $\mu(\Gamma) = 0$ for all $\Gamma \in \mathcal{C}$ satisfying $\nu(\Gamma) = 0$.

ii) Given any measures μ and ν on $(E, \mathcal{B})$ and a $\mathcal{B}$-measurable $f : E \longrightarrow [0, \infty)$, show that $\mu(\Gamma) = \int_\Gamma f\, d\nu$ for all $\Gamma \in \mathcal{B}$ implies that, for every $\Gamma \in \mathcal{C}$, $f \equiv \frac{\mu(\Gamma)}{\nu(\Gamma)}$ on Γ if $\nu(\Gamma) \in (0, \infty]$.

iii) Using the preceding, show that, in general, one cannot dispense with the assumption in Theorem VI.1.13 that ν is σ-finite.

VI.2.15 Exercise.

The readers with good memories may be disturbed by the apparent difference between the notions of absolute continuity used here and that used in Exercise III.3.15 earlier. To allay such concerns, check that, so long as μ is finite, $\mu << \nu$ implies that for every $\epsilon > 0$ there is a $\delta > 0$ with the property that $\mu(\Gamma) < \epsilon$ whenever $\Gamma \in \mathcal{B}$ and $\nu(\Gamma) < \delta$.

VI.2.16 Exercise.

The purpose of this exercise is to take a closer look at the notion of absolute continuity when $E = \mathbb{R}$, $\mathcal{B} = \mathcal{B}_{\mathbb{R}}$, and $\nu = \lambda_{\mathbb{R}}$. In particular, we want to examine this relationship in terms of the mapping $\psi \longmapsto \mu_\psi$ (established in Exercise VI.1.26) taking a bounded, right-continuous, non-decreasing ψ on $\mathbb{R}$ into a finite measure μ_ψ on $(\mathbb{R}, \mathcal{B}_{\mathbb{R}})$.

i) Show that $\mu_\psi << \lambda_{\mathbb{R}}$ if and only if for every $\epsilon > 0$ there is a $\delta > 0$ such that

$$\sum_{m=1}^{\infty} \big(\psi(b_m) - \psi(a_m)\big) \leq \epsilon$$

whenever $\{(a_m, b_m)\}_1^\infty$ is a sequence of open intervals satisfying $\sum_{m=1}^{\infty}(b_m - a_m) < \delta$. Conclude, in particular, that $\mu_\psi << \lambda_{\mathbb{R}}$ only if ψ is continuous.

ii) The preceding makes it reasonable to ask whether there are any bounded, continuous, non-decreasing $\psi : \mathbb{R} \longrightarrow \mathbb{R}$ for which the corresponding μ_ψ fails to be absolutely continuous with respect to $\lambda_{\mathbb{R}}$. In order to show that there are such ψ's, we will now describe the CANTOR–LEBESGUE function.

Referring to Exercise II.1.20, recall that the closed set C_k is the union of 2^k disjoint closed intervals and that $[0, 1] \setminus C_k$ is the union of $2^k - 1$ disjoint open intervals $(a_{k,j}, b_{k,j})$, $1 \leq j \leq 2^k$, where we have ordered these so that $b_{k,j} < a_{k,j+1}$ for $1 \leq j < 2^k$. Next, set $a_{k,0} = -\infty$, $b_{k,0} = 0$, $a_{k,2^k+1} = 1$, $b_{k,2^k+1} = \infty$, and define $\psi_k : \mathbb{R} \longrightarrow [0, 1]$ so that:

a) ψ_k is constant on each of the intervals $[a_{k,j}, b_{k,j}]$, $0 \leq j \leq 2^k + 1$;

b) for each $0 \leq j < 2^k$, ψ_k is linear on each of the intervals $[b_{k,j}, a_{k,j+1}]$ and $\psi_k(b_{k,j}) = j/2^k$.

Notice that each ψ_k is continuous and non-decreasing from $\mathbb{R}$ onto $[0, 1]$. In addition, check that $\|\psi_{k+1} - \psi_k\|_{u,\mathbb{R}} \leq \frac{1}{2^k}$; and conclude from this that ψ_k converges uniformly on $\mathbb{R}$ to a continuous, non-decreasing $\psi : \mathbb{R} \longrightarrow [0, 1]$ with the property that $\mu_\psi(C) = 1$. Since $\lambda_{\mathbb{R}}(C) = 0$, it follows not only that μ_ψ fails to be absolutely continuous with respect to $\lambda_{\mathbb{R}}$ but also that it is even singular to $\lambda_{\mathbb{R}}$!

INDEX